# CARBON MONOXIDE POISONING

# CARBON MONOXIDE POISONING

By

**PROF. DR. MED. K. K. JAIN**
*FRCS(C), FRACS, FICS*
*Neurosurgeon,*
*Consultant in Hyperbaric Medicine*
*Engelberg, Switzerland*

**WARREN H. GREEN, INC.**

***St. Louis, Missouri, U.S.A.***

*Published by*

WARREN H. GREEN, INC.
8356 Olive Boulevard
St. Louis, Missouri 63132, U.S.A.



ISBN No. 87527-258-4

(314) 991-1335

*Printed in the United States of America*

*TO*

*My wife Verena and My children*
*Eric, Adrian, and Vivien for*
*their patience and understanding*
*during preparation of this work*

# PREFACE

Carbon monoxide is a prevalent but underestimated danger to human health. My interest in carbon monoxide poisoning started with the study of neuropathology of cerebral anoxia. Later I studied the role of hyperbaric oxygen in the treatment of carbon monoxide poisoning. Finally, as any other responsible citizen, I am concerned with environmental pollution and public health hazards. Carbon monoxide is an important pollutant in the atmosphere of our cities and the major contributor is the automobile exhaust.

Carbon monoxide poisoning is well known for its role in suicides. Chronic low grade poisoning with carbon monoxide without clear cut clinical manifestations is not well recognized. The commonest cause is chronic heavy smoking. Over one-fourth of the adult population smokes cigarettes which raises carboxyhemoglobin levels. The main concern about tobacco has been lung cancer due to the tar content but less emphasis has been placed on the CO content of the tobacco smoke and its deleterious effects on the human body. Smoking cessation is the most important preventive measure to reduce this source of carbon monoxide poisoning.

This book is written primarily for the physicians as a guide to the understanding and management of carbon monoxide poisoning. There are over 3000 publications on this subject including classical monographs. A comprehensive, handy, and up-to-date source of information on this subject has been lacking. I hope that this book will meet this need. About 600 selected references are distributed at the end of various chapters as a guide for those who want to study the subject further.

Engelberg, Switzerland

K.K. Jain

# CONTENTS


Preface ........ vii

Chapters

1. History of Carbon Monoxide Poisoning ........ 1
2. Biochemical, Physical and Physiological Aspects of CO Poisoning ........ 20
3. Environmental CO and Epidemiology of CO Poisoning ........ 44
4. Smoking and CO Poisoning ........ 65
5. Pathophysiology of CO Poisoning ........ 83
6. Pathology of CO Poisoning ........ 104
7. Clinical features of CO Poisoning ........ 113
8. Diagnostic Procedures for CO Poisoning ........ 128
9. General Principles of Management of CO Poisoning ........ 138
10. Treatment of CO Poisoning by Hyperbaric Oxygenation ........ 147
11. Prevention of CO Poisoning ........ 164
12. Conclusions and Directions for Future Research ........ 174

Index ........ 176

# CARBON MONOXIDE POISONING

# Chapter 1

## HISTORY OF CARBON MONOXIDE POISONING

Human beings have inhaled carbon monoxide (CO) since they have made fire. Open fires inside sheltered caves may have lead to CO poisoning. Some benchmarks in the history of CO poisoning are listed in Table 1-1.

TABLE 1-1
BENCHMARKS IN THE HISTORY OF CARBON MONOXIDE (CO) POISONING

| | |
|---|---|
| Aristotle, 3rd century B.C. | "Coal fumes lead to heavy head and death". |
| Cicero (106-43 B.C.), Rome | Coal fumes were used for suicide and execution. |
| Paracelsus (1493-1541) | Wrote the first treatise on diseases of miners. |
| Von Helmont (1577-1644) | Experimented on himself with "woodgas" from a pot of charcoal and nearly died. |
| Rammazzini (1633-1714) | Wrote De Morbis Arteficum (Diseases of miners). Pointed out the danger of gases from burning coal. |
| Clayton (1688) | Distilled coal gas from coal. |
| Priestley (1772) | Described a combustible gas that burned with a bright blue flame (? carbon monoxide). |
| Harmant (1775), France | First clinical description of coal gas poisoning. |
| Murdock (1792), England<br>1794, Prussia | Proposed the use of coal gas for illumination.<br>First regulations for protection against coal gas poisoning. |
| Cruickshank (1800) | Showed that carbon monoxide is an oxide which can be converted to $CO_2$ by exploding it with oxygen. |
| LeBlanc (1842) | Identified CO as the toxic substance in coal gas. |
| Chenot (1854) | First explanation of the mode of action of CO. |
| Claude Bernard (1857), France | Showed that CO produces hypoxia by reversible combination with hemoglobin. |
| Hoppe (1857), Germany | Showed that CO changes the color of blood to bright red. |
| Linas & Limousin (1868) | First to try oxygen therapy for CO poisoning. |
| Haldane (1895) | Showed that rats survived CO poisoning when placed oxygen at 2 atmospheres pressure. |
| Saint-Martin & Nicloux(1898) | First demonstration of endogenous CO. |
| Mosso (1901) | Suggested the use of hyperbaric oxygen in CO poisoning. |
| Warburg (1926) | CO shown to depress respiratory chain enzymes. |
| End and Long (1942) | Treated CO poisoning in experimental animals by HBO. |
| Migeote (1949) | Detection of CO in atmosphere |
| Smith and Sharp (1960) | First clinical use of HBO in CO poisoning |

Regulations for protection against CO poisoning existed in the Prussian State Laws of 1794. In paragraph 731, it is stated that (translated from German): "Careless burning of coal in closed rooms, the fumes of which can be dangerous to sensitive persons, is, even when no harm has occurred, punishable by 10 Thaler (unit of money of those days) or voluntary imprisonment."

Claude Bernard (1857) was the first to point out that CO produces hypoxia by its reversible combination with hemoglobin to form carboxyhemoglobin. Haldane (1895) showed that poisonous action of CO is entirely due to its power of combining with Hb of the RBC, thus putting them out of action as oxygen carriers. He also showed that the toxic action of CO is eliminated under 2 atmospheres pressure in an oxygen environment because the oxygen-carrying function of the Hb is dispensed with.

Although the explanation given by Haldane is not quite adequate and we consider the effect of CO on the cellular enzyme system, his work is a classic and is reprinted in the appendix to this chapter.

The next most important step in the management of CO poisoning was the experimental work of End and Long (1942) in demonstrating the effectiveness of hyperbaric oxygen (HBO) therapy. Smith and Sharp in Scotland were the first to treat human cases of CO poisoning successfully with HBO.

There have been a few thousand publications on various aspects of CO during the past century. The important monographs and books are listed in Table 2-1. These are also the source for most of the historical information on this subject.

## REFERENCES

End E, Long CW (1942) Oxygen under pressure in carbon monoxide poisoning.
J. Indust Toxicol 24:302-306.

Harmant (1775) cited in, Larcan A (1968) Description de l'intoxication oxycarbonée par D. B. Harmant, de Nancy, en 1775.
Ann Med Nancy 7:169-179.

Hoppe F (1857) Uber die Einwirkung die Kohlenoxidgases auf des Hämatoglobin.
Virchows Archiv Path Anat Physiol Klin Med 11:288-289.

Linas AJ, Limousin S (1868)
Bull Mém Soc Thérap 2:32.

Migeotte MV (1949) The fundamental bond of carbon monoxide at 4.7 u in the solar spectrum.
Phys Ref 75:1108-1109

Mosso A (1901) La mort apparente du coer secours l'empoisonneement par l'oxyd de carboneé.
Arch Ital Biol 35: 75-89.

Nicloux M (1898) sur l'oxyd de carbone contnu normalement dans le sang.
CR Acad Sci (Paris) 126:1526-1528.

Saint-Martin L (1898) Sur le dosage de petites quanitités d'oxyd de carbone dans l'air et dans le sang normal.
C.R. Acad Sci (Paris) 126:1036-1039.

TABLE 1-2
CLASSICAL BOOKS AND MONOGRAPHS ON CARBON MONOXIDE POISONING

| *Author and Year* | *Title and Publisher* | *Comments* |
|---|---|---|
| C. Bernard (1857) | Lecons sur les effects des substances toxiques et medicamenteuses. J. B. Bailliere et fils, Paris. | Rare book, difficult to obtain outside France. Never translated. |
| L. Lewin (1920) | Die Kohlenoxidvergiftung, Springer, Berlin. | Excellent source of history of CO. |
| C. K. Drinker (1938) | Carbon Monoxide Asphyxia, Oxford University Press, New York. | |
| N. Bokonjic (1963) | Stagnant Anoxia and Carbon Monoxide Poisoning. Suppl. 21 Electroencephalography & Clinical Neurophysiology | Exhaustive study of the electrophysiology of CO hyoxia. |
| M. E. Sluitjer (1963) | The treatment of carbon monoxide poisoning by administration of oxygen at high pressure. Doctoral dissertation University of Amsterdam. Once published and distributed in the USA by Charles C. Thomas, Springfield. | Classic work on this subject. Out of print. Clinical information reproduced in Bour and Ledingham (1967). |
| H. Bour and IM Ledingham (1967) | Editors of "Carbon Monoxide Poisoning," Elsevier, Amsterdam-New York. | Containing classic contributions by several authors. |
| I.D. Kim, D.R. Yun (1967) | Carbon Monoxide Poisoning, Seoul, New Medical publication. | Only in Korean. Review of literature plus the largest series of cases of CO poisoning treated by hyperbaric oxygen. |
| WHO (1979) | Carbon Monoxide. | Small paperback but excellent review of the subject. No discussion of treatment. |
| D. Pankow (1981) | Toxikologie des Kohlenmonoxids. VEB Verlag und Gesundheit, Berlin (East Germany). | Contains 1987 references to literature. Excellent review of the basic aspects. No discussion of the treatment by HBO. |
| R.J. Shepherd (1983) | Carbon Monoxide: the silent killer. Springfield, Charles C. Thomas. | Excellent description of the biological, epidemiological, and medico-legal aspects of the subject. Does not include treatment. |

## APPENDIX*

# THE RELATION OF THE ACTION OF CARBONIC OXIDE TO OXYGEN TENSION.

By John Haldane, M.A., M.D., *Lecturer in Physiology, University of Oxford. Grocers' Company Research Scholar. (Two Figures in Text.)*

*[From the Physiological Laboratory, Oxford.]*

The following investigation took its origin from an enquiry in which I am at present engaged into the nature and action of the suffocative or poisonous gases met with in the air of coal-mines. Among these gases is carbonic oxide. So far as my experience goes, however, carbonic oxide never occurs alone as an impurity, but always in connection with an excess of diluent gases, chiefly nitrogen; so that when carbonic oxide is present in the air of a coal-mine, there is also a more or less considerable reduction in the percentage of oxygen. It is known, however, firstly, that part at least of the action of carbonic oxide is due to its property of entering into combination with the hæmoglobin of the blood corpuscles, and so putting them out of action as oxygen carriers; and, secondly, that the proportion of carbonic-oxide hæmoglobin formed in blood brought into contact with a gas-mixture containing carbonic oxide and oxygen, depends not merely on the tension of carbonic oxide in the mixture, but also on the tension of oxygen. The larger the proportion of oxygen in the gas mixture the smaller will be the amount of carbonic-oxide-hæmoglobin formed in the presence of a given proportion of carbonic oxide. Hence it might be expected that in air containing a diminished proportion of oxygen, carbonic oxide will be more poisonous; and that in air containing an increased proportion the poisonous action will be less. A very simple experiment served to show that this is the case.

Two bottles, each of about three litres capacity, were filled, one with oxygen, and the other with air. Into each bottle 15 c.c. of carbonic oxide were introduced through a tubulated stopper. The bottles were then closed, and the gases mixed by thorough shaking with a little water which had been left for the purpose inside. The water having been removed a young mouse was introduced into each of the bottles, which were again closed. Within five minutes the mouse in the bottle containing air showed marked signs of loss of power over the limbs, and two minutes later it was lying on its back comatose. Fresh air was then at once blown through the bottle with a bellows. Two minutes later the mouse had regained power over its limbs, and about twelve minutes later it seemed quite in its normal condition again. Meanwhile the mouse in the bottle containing oxygen remained quite unaffected. After an hour and a half it was taken out. For some time previously it had been shivering, and was evidently suffering from cold, as it had got rather wet in the bottle, but it showed no symptom of poisoning. Both mice remained perfectly well when replaced in their cage.

It seemed of importance to determine as accurately as possible within what limits raising or lowering of the oxygen percentage affects the poisonous action

* Reprinted from J Physiology (London) 18 (1895):201-217. by permission.

of carbonic oxide. For this purpose the following arrangement was employed (Fig. 1).

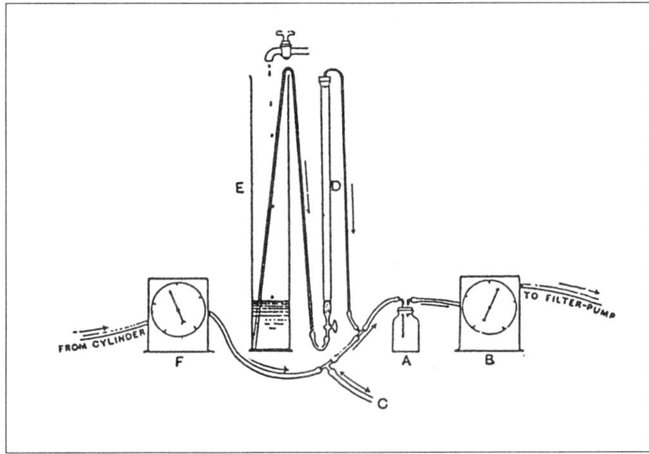


Fig. 1-1

A mouse was placed in the bottle *A*, which had a capacity of about 200 c.c.: through this bottle a perfectly steady air-current in the direction shown by the arrows was maintained by means of a filter pump driven by water supplied at a constant pressure from a cistern. The air current was measured by the small meter *B* (the accuracy of which was ascertained) and was drawn into the tubing at the opening *C*. Between *C* and *A* a T tube was placed, communicating by one of its limbs with narrow glass tubing, which was connected in the manner shown by means of a paraffined cork with the upper end of a burette *D*, containing carbonic oxide. The nozzle of the burette was connected with a siphon, dipping into a tall glass jar *E*. Over this jar was a water tap. When water was allowed to drop from the tap into the jar the level of the water in the latter, and in the burette gradually rose, and a measured quantity of carbonic oxide could thus be steadily and slowly driven from the burette into the air current. By regulating the water-tap the rate of outflow of carbonic oxide could be regulated at will. The water-tap was not connected directly with the burette because a steady outflow of water could not be obtained unless the rate of flow was considerable.

When it was desired to substitute oxygen or any other gas for the whole or part of the air the oxygen was allowed to pass in from a cylinder of compressed gas through the meter *F*. When pure oxygen was required the flow was so regulated that a slight excess of oxygen passed outwards by the opening *C*. By means of this arrangement it was possible rapidly to substitute oxygen for air without the slightest interruption or alteration in the rate of flow through the bottle and meter *B*.

The carbonic oxide employed was made in fairly large quantities at a time by heating oxalic acid with sulphuric acid, and kept in a large bottle fitted up as a gasholder. The gas was purified by passing it through a large tube of moist potash-lime. Each new stock of carbonic oxide was accurately analysed to ascertain the degree of purity of the gas. Sometimes, as in the example given below, it was found that, in spite of care, appreciable quantities of air had become mixed with the carbonic oxide. In calculating the percentages allowance was in each case made for the impurity. The oxygen was determined with pyrogallic acid and potash, and the carbonic oxide by ammoniacal solution of cuprous chloride. The apparatus was the same as that which I have used in the analysis of gases from coal-mines.[1]

*Example of an analysis.*

| | | |
|---|---|---|
| Carbonic oxide | 92·27 | |
| Carbonic acid | 0·07 | |
| Oxygen | 1·56 | |
| Nitrogen | 6·10 | } = 7·66 of air. |
| | 100·00 | |

A month later, after nearly the whole of this bottleful of gas had been used, the residue was found to contain 0·15 % of carbonic acid and 1·96% of oxygen (corresponding to 9·4 % of air), so that there was only a very slight additional contamination of the gas by air diffusing from the water.

As good reasons exist for suspecting that the gas employed in some of the recorded experiments on carbonic oxide was impure a contained carbonic acid or air, it seemed necessary to take every precaution as regards the gas employed.

In air about 0·06% of carbonic oxide is sufficient to produce distinct symptoms in mice. The following are the notes of an experiment.

Young black and white mouse about half grown. Rate of ventilation 1·70 litres per minute. Rate of CO delivery 1·00 c.c. per minute. Corrected percentage of CO= ·058%[2]. Temperature 13° C.

| | |
|---|---|
| 5 p.m. | Mouse placed in bottle and left in the air current, no CO being added. |
| 5.38 | No change. Quite active. Movements as sharp as usual. Climbs up glass tube, cleans its fur, &c. as usual. CO now added. |
| 5.42 | Slight hyperpnoea, the respiratory movements being more visible. |
| 5.47 | Mouse less active. Sits up on hind legs, panting slightly. Does not clean its fur, &c. as usual. |
| 5.51 | CO off. |

[1] ***Proc. Roy. Soc.* 57, p. 244,** 1895, and *Transactions of the Federated Institution of Mining Engineers*, 1895.

[2] The stock of carbonic oxide used in this experiment had the following composition:

| | | |
|---|---|---|
| Carbonic oxide | 97·42 | |
| Carbonic acid | 0·05 | |
| Oxygen | 0·48 | } air 2·53. |
| Nitrogen | 2·05 | |

5.57 Cleaning its fur vigorously, climbing up the tube, and is altogether much more lively. No longer panting, and appears to be quite as usual again.
5.59 CO again started.
6.4 Panting distinctly. On trying to climb up it gets exhausted, and sits panting.
6.13 Remains very quiet. Sitting in one position, and apparently panting.
6.24 Seems drowsy and torpid. When roused so that it tries to climb the glass a tendency to slip down is apparent, so that there is some loss of power in the limbs.
6.30 Bottle containing mouse was put in warm water at about 35°CO.
6.35 Panting much less, but torpor and feebleness remain.
6.36 Bottle put in cold water at 12°. Panting began again in about minute.
6.55 After the last observation had been repeated several times the bottle was placed in warm water at 26° and left.
7.6 Quite as torpid and feeble as ever. CO off.
7.12 Had regained its full power over limbs, climbed up tube without exhaustion, and then began to clean its fur.
7.15 Quite as lively as usual again.

The following experiment will afford an idea of the effects of a somewhat higher percentage of carbonic oxide.

Young mouse, not quite fully grown. Rate of ventilation 1·03 litres per minute. Rate of CO delivery 2·34 c.c. per minute. Corrected percentage of CO 0·221%. Temperature 14·5°.

10.55 CO started.
10.59 Hyperpnœa and great loss of power over legs. Cannot stand. Lies on belly with legs stretched out. Seems hardly conscious.
11.12 Continues very helpless and drowsy looking. Remains lying on side when put there, but does not remain on its back.
11.20 Same condition, but respirations seem less frequent (108). [Normal frequency = about 140.]
11.35 Respirations 90. Seems feebler.
11.46 Respirations 74 and are feebler. Remains on back when put there.
11.57 Respirations 66. Has lain on its back since last observation.
12.8 Respirations 44.
12.27 Respirations 35.
12.49 Respirations 16 and gasping in character.
1.20 Died almost imperceptibly. Since last observation respirations had gradually become shallower, until at last they died away.

On post mortem examination the liver and other organs had the characteristic bright red appearance seen in carbonic oxide poisoning. On diluting a little of the blood with water the colour of the solution indicated that the blood was about two thirds or three-fourths saturated with CO.

With larger proportions of carbonic oxide than in the last experiment death is correspondingly more rapid, and is accompanied by convulsions, as in the case of asphyxia from breathing an atmosphere deprived of oxygen. If death occurs within a few seconds the blood in some parts of the body may, as Heger has already observed[1], contain very little carbonic oxide hæmoglobin, death being too rapid to allow the whole of the blood to become saturated.

The observation that so small a proportion 0·06 % of carbonic oxide produces distinct symptoms is not new, but has previously been made by Hempel[2].

It is probable that, particularly with such small animals as mice, death by poisoning with very small proportions of carbonic oxide is due partly to fall of the body temperature consequent on diminished metabolism and heat production. The following experiment, in which the production of carbonic acid by a mouse was estimated, supports this view.

The apparatus was the same as that already described (Fig. 1) except the air entering the bottle was deprived of carbonic acid by soda lime, while the carbonic acid in the outgoing air current was estimated by means of absorption apparatus described by Mr. Pembrey and myself[3].

| *Time* | | *$CO_2$ given off in grammes* | *Air breathed by mouse* |
|---|---|---|---|
| First | 15 minutes | ·0552 | Pure air |
| Second | " " | ·0570 | Pure air |
| Third | " " | ·0531 | 0·16% of CO |
| Fourth | " " | ·0264 | 0·16% of CO |
| Fifth | " " | ·0272 | 0·16% of CO |
| Sixth | " " | ·0227 | Pure air |
| Seventh | " " | ·0405 | Pure air |
| Eighth | " " | ·0508 | Pure air |

The carbonic oxide produced the characteristic loss of power over the limbs, &c. Recovery was rather slow. During the seventh and eighth periods shivering was observed, as if the animal were suffering from Probably its temperature had gone down under the influence of the carbonic oxide. The bottle containing the animal was kept in a bath at 15·5° C, and the rate of ventilation was 0·48 litres per minute.

It will be seen that the carbonic oxide produced a very marked diminution in the respiratory exchange, and that the diminution lasted for about half-an-hour after the animal had ceased to breathe carbonic oxide.

## EXPERIMENTS WITH CARBONIC OXIDE IN OXYGEN

The oxygen used in these experiments was that supplied cylinders by Mr Orchard. It contains about 97% of oxygen. following is the result of an analysis of a sample from the cylinder employed in most of the experiments.

| | |
|---|---|
| Oxygen | 96·91 |
| Carbonic acid | 0·25 |
| Nitrogen | 2·84 |
| | 100·00 |

[1] *Journal de Medecine de Chirurgie et de Pharmacologie*, 1894, p.106.

[2] *Zeitschrift für analytische Chemie*, XVIII. p.399.

[3] *Philosophical Magazine*, 1890, p. 306.

I have breathed this oxygen pure for considerable periods without its producing any appreciable effects. It has no smell suggestive of chlorine or other impurity.

In oxygen it requires about ·8% of carbonic oxide to distinctly affect a mouse, and very much more to produce death. The effect of dangerous percentages (over 5%) seems to vary considerably according to circumstances. When the percentage is gradually increased the mouse appears to be capable of adapting itself to a considerable extent to the atmosphere. The following two experiments may be quoted.

Young mouse, not quite fully grown. Initial ventilation ·28 litres per minute. Bottle kept in bath at 28°.

12.20 Mouse put in bottle and ventilated with oxygen.
12.37 CO started. Corrected percentage = ·85.
12.41 Slight hyperpnœa, and seems a little drowsy.
12.46 Distinct sluggishness, and apparent slight loss of power, noticed when it jumps up on the side of bottle.
12.55 Condition same. Ventilation reduced, so that CO percentage = 1·76.
12.58 Hyperpnœa and marked loss of power. On trying to climb up it stops to rest, or slips down.
1.5 Seems no worse. Lies or sits in awkward positions, as if very drowsy. When roused gets up on legs.
1.14 No worse. Began to clean its fur, but stopped again at once.
1.20 Ventilation increased from ·13 to ·66 litres per minute (i.e. to five times its previous amount). Oxygen turned on and air substituted. Percentage of CO now = ·34 (i.e. a fifth of its former amount.)
1.22 Seems decidedly more feeble, and rapidly losing power.
1.26 Remains lying on its side or back in a comatose state.
1.26 CO exhausted. The mouse had recovered completely a few minutes later.

Same mouse. Ventilation throughout = ·102 litres per minute. In bath at 28°.

2.5 Mouse put in and ventilated with pure oxygen. After a short time went to sleep.
2.12 CO started. Percentage = 2·31.
2.15 No change in the mouse. Sitting asleep.
2.17 Laying head down; seems unnaturally drowsy and losing power over limbs.
2.22 Lies with legs spread out, but will not remain on its back.
2.28 Condition same. CO increased to 2·97%.
2.39 Not apparently more feeble. Still will not lie on its back. Sometimes wakes up and tries to climb.
2.41 CO increased to 7·2%.
2.47 Still does not remain on its back, and seems no worse. CO exhausted.
3.0 Completely recovered. Now cleaning its fur.

These experiments sufficiently show that in oxygen carbonic oxide is very much less poisonous than in air. This fact can also be very conveniently demonstrated with the apparatus described by alternately supplying the animal with air and oxygen, while the delivery carbonic oxide and the rate of ventilation remain absolutely unaltered. If about ·2 to ·5% of carbonic oxide are added to the ventilating current the animal will be seen to alternately become worse or recover again according as air or oxygen is supplied. With a somewhat higher percentage death occurs very rapidly when air is supplied.

## EXPERIMENTS WITH DIMINISHED OXYGEN PERCENTAGES

To obtain an atmosphere containing a diminished percentage of oxygen, hydrogen or nitrogen was partially substituted for air by means of the arrangement already described. The hydrogen employed gave no arsenic reaction with Marsh's test, and portions from the same cylinder had previously been used for experiments on man without any ill effects being produced.

Mice, as compared with men, appear to be specially sensitive to reduced percentages of oxygen. A mixture of one-third of hydrogen or nitrogen and two-thirds air causes panting and uneasiness. Cyanosis and loss of power over the limbs is caused by a mixture containing two-thirds of hydrogen. With considerably reduced percentages of oxygen it is thus difficult to distinguish the symptoms produced simply by lack of oxygen from those due to carbonic oxide. Nevertheless it is quite evident that the poisonous action of carbonic oxide is very mark increased by a diminution of the oxygen percentage. The following experiment may be quoted.

Half grown brown mouse. Ventilation 100 litres per minute. Temperature 12.5°.

1.7 Mouse put into bottle and ventilated with pure air.
1.9 Continues quite lively. Hydrogen now added at the rate of ·33 litres per minute, so that oxygen percentage was reduced from 20·9 to 13·9.
1.12 Mouse panting a little, and is not so lively. Still cleans its fur, &c. but often stops to pant.
1.21 Mouse remains the same. No loss of power. Hydrogen off.
1.24 Panting ceased. Much more lively.
1.25 CO turned on. Percentage = 0·247.
1.27 Panting.
1.29 Limbs getting feeble.
1.32 Limbs sprawling. Mouse sits resting on its belly. Will not lie on side or back.
1.35 No further appreciable change in mouse. Hydrogen turned on at same rate as before.
1.36 Mouse much worse. Lies on back.
1.36 Respirations gasping in character. Mouse lying on back with limbs twitching.
1.38 Seems to be dying. Hydrogen off.
1.39 Improving. Will not now lie on back. Twitching and gasping respirations have disappeared. Limbs still sprawling.
1.40 1/2 Hydrogen on again.
1.41 1/2 Gasping respirations.
1.42 1/2 Remains on back or side. Quite comatose.
1.46 Hydrogen off.

1.51 Remains in same condition. Respirations getting less frequent, CO stopped.
1.58 Still quite comatose. Taken out. Felt very cold, therefore warmed in the hand, when it gradually began to revive, and was replaced in its cage in cotton-wool.

At 2.50 it was still feeble, but when seen again eight hours later it was perfectly well.

A similar very marked increase in the poisonous action of carbonic oxide was observed in several other experiments with hydrogen air: also in an experiment in which a mixture of 31·4% of pure nitrogen and 68·6% of air was employed.

The conclusions drawn from these experiments with carbonic oxide in air containing a reduced percentage of oxygen have of course a special practical interest in connection with carbonic oxide poisoning as it occurs in coal-mines and elsewhere, but I hope to deal more fully with this subject in another paper.

## DISCUSSION OF THE RESULTS

In the case of pure oxygen the tension of oxygen is nearly five time as great as in air, so that to produce equal saturation of the hæmoglobin of the corpuscles with carbonic oxide in oxygen and in the tension of carbonic oxide would presumably require to be five times as great in the oxygen as in the air. Hence on the hypothesis on which the investigation started one might expect that carbonic oxide would turn out to be about five times as poisonous in air as in oxygen that is to say, that five times as high a percentage of carbonic oxide would be required in oxygen to produce the effect of a given percentage in air.

Now the experiments made with the apparatus described above clearly showed that carbonic oxide is much more than five times as poisonous in air as in oxygen. In the latter gas the poisonous action is reduced to about a tenth or less. Evidently then some other factor has to be taken into account besides the relative tensions of oxygen carbonic oxide. That this factor exists was further shown by following experiment:

A thick-walled bottle of 3·1 litres capacity was filled over water with oxygen, to which was added 30 c.c. of carbonic oxide, so that the oxygen contained about 1% of carbonic oxide. After the latter had been thoroughly mixed with the oxygen a full-grown mouse was placed in the bottle, which was again tightly closed by means of a tubulated cork, the tube of which connected with a filter-pump and gauge, so that the pressure in the bottle could be reduced as required. By means of a clamp the bottle could be shut off from the pump when any desired pressure was obtained.

4.40 Mouse put into bottle.
4.42 Some hypernœa.
4.56 No further change. Walks and climbs about normally. No distinct loss of power over limbs, though perhaps a little sluggish.
4.56-57 Pressure reduced to about 50% of an atmosphere.
5.1 Hyperpnœa more marked. More tendency to stop and pant.

5.5 No further change. Pressure reduced to 27% of an atmosphere.
5.8 Panting greater. Looks drowsy and totters when walking.
5.11 Marked weakness of limbs. Tends to sprawl on its belly or lie on its side but can creep about when roused.
5.15 No further change. Pressure reduced to about 17% of an atmosphere.
5.20 No marked change. Pressure reduced to about 10% of an atmosphere.
5.25 Perhaps a little more helpless, but no marked change. Air was now let in to atmospheric pressure, when the mouse rapidly improved, and recovered completely on being taken out.

In this experiment the oxygen tension and carbonic oxide tension were diminished simultaneously and in equal proportions. The effect of the carbonic oxide nevertheless increased.

To account for this relation between oxygen tension and the effect of carbonic oxide the hypothesis suggested itself that the higher the oxygen tension the less dependent an animal is on its red corpuscles as oxygen carriers, since the oxygen simply dissolved in the blood becomes considerable when the oxygen tension is high.

## EXPERIMENTS WITH INCREASED OXYGEN PRESSURE

If the above hypothesis were correct one would expect to find that by raising the oxygen tension sufficiently high it would be possible to abolish entirely the poisonous action of the carbonic oxide. There is, however, a limit to the possibility of raising the oxygen tension, since, as shown by Paul Bert, at about five atmospheres, or somewhat less, oxygen acts as a poison. The question therefore was whether at a less tension than this the action of carbonic oxide could be abolished. To investigate this the following arrangement was employed (Fig 2).

Three thick-walled measuring cylinders, *A*, *B*, *C*, each of about 650 cc. capacity, and provided with tightly fitting paraffined corks, tubulated in the manner shown, were connected together by means of thick-walled rubber tubing. *A* was further connected, in the manner shown, with a cylinder of compressed oxygen, and a mercury pressure gauge *E*, and with the vessel *C*. Screw clamps were placed on the tubing at *E* and *F*, and there was a three-way glass tap at *D*. The joints, corks, etc. were carefully secured so as to withstand the required pressure. The vessel *B*, having been first filled with water, was connected with the gas-holder of carbonic oxide, and filled with carbonic oxide, the water being sucked over into *C*. The tap *D* was then closed. A mouse was now placed in *A*, which had been previously filled with oxygen. The cork having been pressed home and secured, the pressure in the whole system was gradually raised to the required amount by cautiously turning the regulating screw of the oxygen cylinder. The clamp *E* was now closed, and *D* turned so as to connect A with *B*. Carbonic oxide could then be driven from *B* into *A* as required, by means of pressure from the oxygen cylinder.

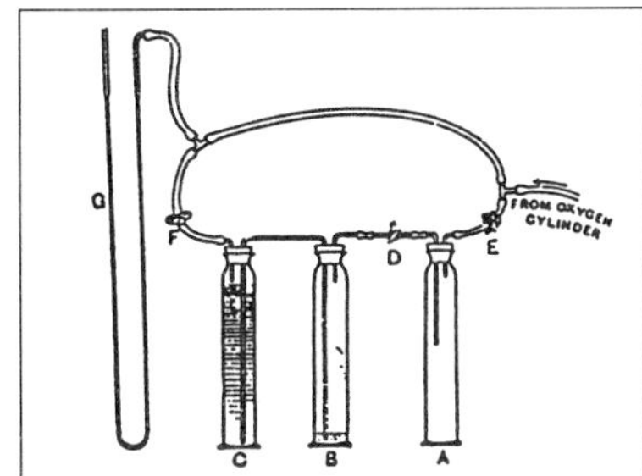


Fig. 1-2

The following are the notes of an experiment.

7.15 Large mouse put in.
7.21 Pressure raised to 70·5 cm. of mercury.
7 26 Mouse unaffected. Apparatus tight. 30 c.c. of CO (giving a tension of 6% of an atmosphere of CO in *A*) driven into *A*.
7.28 Slight hyperpnœa.
7.29 35 c.c. of CO sent in, so that tension of CO in *A* about 15% of an atmosphere.
7.31 Hyperpnœa more marked. Often jumps up in vessel, as if uneasy.
7.32 50 c.c. of CO sent in so that CO tension in *A* = about 27% atmosphere.
7.35 Mouse gets easily exhausted after any effort, such as jumping up in the bottle. 75 c.c. of CO sent in, so that tension of CO in *A* = nearly 50% of an atmosphere.
7.38 Mouse as before; can walk about quite well when *A* is held horizontally, but is exhausted easily by any effort. The whole of the CO now sent in, *D* and F closed and E opened. Pressure in A = 139·5 cm., so that tension of CO in *A* = 139.5 — 70.5 = 69 cm., or nearly an atmosphere.
7.47 No change in mouse. Walks about, and climbs on the tube when *A* is held horizontally, but is easily exhausted and somewhat sluggish.
7.47-49 Pressure in *A* diminished to 63 cm. by opening *D* outwards. Mouse rapidly became helpless and unconscious, and began to gasp. Died at 4.51.

Postmortem. Liver and other organs bright red. A drop of blood largely diluted with water gave the characteristic cherry-red colour, which was hardly appreciably altered on shaking some of the solution with coal-gas. The blood was thus practically saturated with carbonic oxide.

This experiment shows that at a tension of two atmospheres of oxygen the poisonous action of carbonic oxide on mice is abolished. Apart from its action in putting the red corpuscles out of action as oxygen carriers carbonic oxide would thus appear to be a physiologically indifferent gas, like nitrogen[1].

I have several times repeated the experiment just described, but have never found it fail. In some experiments the mouse was kept for half-an-hour or more in the vessel, until the carbonic acid tension must have been nearly great enough to affect the animal.

[1] Experiments interpreted as leading to an opposite conclusion have been recently described by Marcacci (*Archives Italiennes de Biologie XIX*. p. 205,1893) and Piotrowski (*Archiv f. Anat. and Physiol.* 1893, p. 205). Neither of these authors, however, furnishes analyses of the gas employed by them, or any other guarantee of its freedom from such probable impurities as carbonic acid and air.

It might be thought that the death of the mouse is perhaps due to the liberation of gas within its body, from too sudden diminution of pressure. Control experiments with pure oxygen or air showed, however, that this is not the case. Even a very sudden diminution of pressure, from giving way of a cork, did not injure the animal. Diminution of the pressure of a mixture of oxygen with a third of carbonic oxide does not, however, always, kill a mouse, as shown by the following experiment.

A young half-grown mouse had been kept since 1·25 at a pressure of 128 cm. of mercury in an atmosphere consisting of about two-thirds of oxygen and one-third of carbonic oxide. At the end of this time it remained quite lively.

1.49 Pressure reduced to 106 cm.
1.50 Decidedly worse. Legs tend to sprawl.
1.52 Pressure reduced to 78 cm.
1.55 Lying with its head on the bottom of the jar, but does not remain on its back when put there.
1.57 2·8 Pressure gradually reduced to 0.
2.15 Mouse more torpid, but still does not remain on its back. Removed from the glass vessel and brought into the air. Immediately got worse. Lay on its back and had slight convulsions. Breathing ceased at 2.16 except for occasional gasps. Appeared once or twice to be dead.
2.18 Respirations becoming more frequent.
2.23 Still unconscious and does not move limbs.
2.27 Shows signs of returning consciousness, and power over its limbs.
4 p.m. Completely recovered.

Young mice are probably more capable of resisting asphyxia from carbonic oxide poisoning than older ones. This is what might be expected from the fact that young animals are difficult to drown, and may live for a short time in an atmosphere deprived of oxygen.

To remove a mouse with safety from the atmosphere of carbonic oxide in the pressure apparatus it is necessary to replace the carbonic oxide by oxygen without in the operation letting the pressure down to less than about two atmospheres. When the carbonic oxide has been washed out the animal can be safely removed to air.

In support of the hypothesis that carbonic oxide is, apart from its influence on hæmoglobin, a physiologically indifferent gas I may quote here the following experiment, made on an animal devoid of hæmoglobin.

Two mixtures are made, the one of 20 volumes of oxygen and 80 of carbonic oxide, the other of 20 volumes of oxygen and 80 of carbonic acid. Into each of these mixtures a cockroach (Blatta orientalis) is brought. In the carbonic oxide mixture the animal is not sensibly affected, even after a week [1]. In the carbonic acid mixture, on the other hand, a cockroach almost instantly exhibits convulsive movements, and becomes quite motionless at the end of from 20 to 30 seconds[2]. It appears to be dead, but nevertheless recovers after a time if taken out before too long. This experiment shows in a striking way the contrast between the essentially indifferent gas, carbonic oxide, and the essentially poisonous gas, carbonic acid.

It is now necessary to discuss more in detail the probability of the hypothesis advanced above that the abolition of the poisonous action of carbonic oxide when the oxygen tension is raised to two atmospheres is due to the fact that the animal can live on the oxygen simply dissolved in the blood.

In the absence of directly obtained data as to the coefficient of absorption of oxygen in blood we may provisionally take as a basis for calculation the number ·0262, given by Zuntz[3] as probable from the results of experiments by Paul Bert on the solubility of nitrogen in blood. Arterial blood contains usually about 20% of its volume of oxygen, combined in the red corpuscles. Of this oxygen about six or eight volumes are, under ordinary circumstances, used up in the circulation, so that six or eight volumes of oxygen would appear to be necessary to supply the normal requirements of the tissues. Now on the above assumption blood will retain in simple solution about 2·6% of its volume of oxygen in presence of an atmosphere of pure oxygen, or about 0·5% in presence of air. At two atmospheres of oxygen blood will dissolve 5·2% of oxygen. This dissolved oxygen will, moreover, probably be particularly easily available for the wants of the tissues, since its tension is very high, and it will therefore probably pass very easily and completely through the capillary walls.

Now 5·2 volumes percent of oxygen will be scarcely as much as is required by the animal. Nevertheless the latter possesses an easy means of making this reduced quantity suffice—namely by increasing the rate of the circulation. During any serious exertion the respiratory exchange of such animals as have been investigated is enormously increased—often to about ten times the normal value, or probably more. To cover this increase the rate of circulation must apparently be also very largely increased, probably by several times. Even were the whole of the oxygen of the arterial blood used up the quantity would not be anything like

---

[1] In one experiment a cockroach was kept without injury in a carbonic oxide mixture for 18 days. At the end of this time the residual atmosphere had the following composition:

| | |
|---|---|
| Carbonic oxide | 71.41 |
| Oxygen | 14.21 |
| Nitrogen | 9.34 |
| Carbonic acid | 5.04 |
| | 100.00 |

[2] An almost equally rapid effect is produced by putting the animal in pure hydrogen, and thus depriving it of oxygen.

[3] *Hermann's Handbuch,* IV. 2, p. 16. .

sufficient to supply the needs of the tissues during exertion, unless the rate of circulation were largely increased.

If then the rate of circulation were similarly increased in an animal deprived of the use of its red corpuscles, but breathing in an atmosphere of oxygen at double the normal pressure, the requirements of the tissues for oxygen could easily be covered. It is also possible to see how, as in the last experiment, an animal might still live for a short time when deprived of the use of its corpuscles in an atmosphere at ordinary pressure consisting of two parts of oxygen and one of carbonic oxide.

There can be little doubt that with high carbonic oxide tensions such as those employed in the pressure experiments just described, the corpuscles are practically saturated with carbonic oxide, and so put completely out of action [1]. Both theoretical considerations, and the examination of the blood of various animals which have died on lowering the pressure, confirm this assumption. In a future paper I hope, however, to communicate more direct evidence on this and a number of further points connected with, and suggested by, the action of carbonic oxide on mice and other animals, including man.

The theory advanced seems to explain completely both the abolition, at high oxygen tensions, of the poisonous action, and The effect of simultaneously diminishing the oxygen and carbonic oxide tension. As might be expected the effect of simultaneous diminution is most striking with high initial oxygen tensions. With initial tensions of less than an atmosphere the oxygen simply dissolved in the blood is of less relative importance to the animal, so that the symptoms come to depend chiefly on the relative affinities for haemoglobin of the two gases and the effect on the animal of simultaneous proportional diminution of their tensions is less marked, as in the experiment described on page 209. Finally, the theory explains the fact that the slight effects produced by carbonic oxide even in presence of oxygen at two atmospheres of pressure do not increase on raising the carbonic oxide tension above about 10% of an atmosphere.

---

[1] It might be thought that possibly the relative affinities of oxygen and carbonic oxide for haemoglobin are different at high pressures from what they are at low pressures, and that consequently the corpuscles are not saturated with carbonic oxide at the pressure employed in these experiments. To test this hypothesis I put a dilute solution of blood in the pressure apparatus, with a pressure of two atmospheres of oxygen, and observed its changes of colour on letting in carbonic oxide. But even with a tension of as little as 10% of an atmosphere of carbonic oxide the solution gave on shaking the same tint as part of the same solution saturated with carbonic oxide. The saturation of the blood was thus practically complete at this tension, in spite of the high oxygen pressure.

## CHIEF CONCLUSIONS

1. The poisonous action of carbonic oxide diminishes as the oxygen tension increases, and *vice versa.* At a tension of two atmospheres of oxygen this poisonous action is abolished in the case of mice.

2. The disappearance of the poisonous action is due to the fact that at high oxygen tensions the animals can dispense entirely with the oxygen-carrying function of haemoglobin.

3. The poisonous action of carbonic oxide is entirely due to its power of combining with the hemoglobin of the red corpuscles, and so putting them out of action as oxygen-carriers.

# Chapter 2

## PHYSICAL, BIOCHEMICAL, AND PHYSIOLOGICAL ASPECTS OF CO POISONING

### PHYSICAL AND CHEMICAL PROPERTIES

Carbon monoxide is a colorless, odorless and tasteless gas. It is not combustible alone but burns with a blue flame in the presence of oxygen. It is less soluble in water than other gases but is markedly soluble in organic solvents such as chloroform and acetic acid. It is 7 times more soluble in alcohol than in water. Physical properties of CO are shown in Table 2-1. CO is formed during incomplete combustion of carbonaceous material in the environments. It absorbs electromagnetic radiation in the infrared region with the main absorption band centered at 4.67 µm; this property is used for the measurement of CO concentrations in air.

TABLE 2-1
PHYSICAL PROPERTIES OF CARBON MONOXIDE

| | | |
|---|---|---|
| Molecular weight | | 28.01 |
| Melting point | | -205.1° C |
| Boiling point | | -191.5° C |
| Density | at 0° C, 1 atm | 1.250 gm/litre |
| | at 25° C, 1 atm | 1.145 gm/litre |
| Specific gravity relative to air | | 0.967 |
| Solubility in water | at 0° C, 1 atm | 3.54 ml/100 ml |
| | at 25° C, 1 atm | 2.14 ml/100 ml |
| | at 37° C, 1 atm | 1.83 ml/100 ml |
| Conversion factors | | |
| | at 0° C, 1 atm | 1 mg/m$^3$ = 0.800 ppm |
| | | 1 ppm = 1.250 mg/m$^3$ |
| | at 25° C, 1 atm | 1 mg/m$^3$ = 0.873 ppm |
| | | 1 ppm + 1.145 mg/m$^3$ |

1 atm = pressure of atmosphere at sea level

ppm = parts per million by volume

CO competes with oxygen for binding to mammalian proteins but its structure resembles more that of the inert gas nitrogen than that of oxygen. The arrangement of electrons in the formal molecular orbital structure of CO is the same as in nitrogen where all occupied orbitals are shared by electron pairs.

$$:N{\equiv}N: \quad :C{\equiv}O:$$

CO is isoelectric with nitrogen, nitrosyl, and cyanide ions. The restitution of the electronic structure of CO by Lewisonian electron formula is controversial. Here the element symbols denote the atomic nucleus plus the electron orbits with the exception of valency orbits whose electrons are shown by dots. The following borderline states of CO exist:

$$:C^{\oplus}:\ddot{\underset{..}{O}}^{\theta}: \qquad C::\ddot{O}: \qquad :C::\underset{..}{O}: \qquad :C^{\theta}:::O^{\theta}:$$

According to dipole moment and resonance energy, these borderline states occur with a frequency of 10, 20, and 50 percent respectively.

CO is an anhydride of formic acid. Formates are formed when CO is heated with alkalies. Boudeward equation is:

$$2CO_2\ C + CO_2$$

It describes the breakdown of CO into $CO_2$ and carbon which predominates in the temperature range of 400-700°C. Over 800°C the equation swings in favor of CO formation.

CO and hydrogen mixtures are also formed as a catalytic and thermal reaction from methane, ethane, propane, and butane:

$$CH_4 + H_2O \rightarrow CO + 3\,H_2 \text{ or } CH_4 + 1/2\,O_2 \rightarrow CO + 2\,H_2$$

In the laboratory, CO can be produced from formic acid and sulfuric acid:

$$HCOOH \rightarrow CO + H_2O$$

or from heating dry calcium carbonate with powdered zinc:

$$CaCO_2 + Zn \rightarrow CaO + ZnO + CO$$

CO can be converted to $CO_2$ by interaction with nitric oxide in the presence of rhodium (Rh) catalyst (Oh *et al.* 1986).

$$2\,NO + 2\,CO \underset{Rh}{\rightarrow} 2\,CO_2 + N_2$$

This reaction is important in the control of atmospheric pollution with CO from automobile exhausts as modern automotive catalytic converters contain Rh.

Although CO is a neutral diatomic molecule, it will serve as a ligand to metals to form metal carbonyls. Because of electron dispersion, CO can serve as an electron donor and a pair of electrons for C-atoms can be shared with a metal atom. Perturbations of electronic structure around the metal atom can be monitored in terms of stretch frequencies and infrared spectra. CO has been shown to bind the carbon atom towards the metal atom in a bonding back-bonding scheme (Paul *et al.* 1985). Metal-CO binding depends not only on the presence of a suitable metal but also on the availability of a binding site (Caughey 1970). CO tends to form stable linear metal ligand complexes with various metals such as iron. CO combines with ferrohemoglobin and ferromyoglobin and interferes with oxygen transfer. Binding of CO with heme proteins has been studied by x-ray absorption near edge structural spectroscopy to determine the variation of CO-bond angles (Bianconi *et al.* 1985). CO binds to heme proteins in a bent configuration. The linear bonding of CO is prevented mainly by steric hindrance of the distal histid-

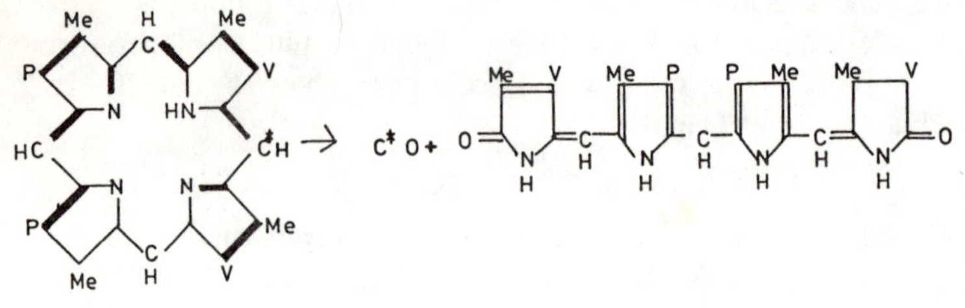

PROTOPORPHYRIN → C* O + BILIRUBIN

Fig. 2-1: Catabolism of heme (a protoporphyrin) giving rise to one mole of endogenous CO and bilirubin.

ine. CO must fit into a tight pocket designed to bind oxygen in a bent configuration.

Although the chemistries of CO and oxygen are markedly different, both require the same oxidation states of either $Fe^{+}$ or $Cu^{1+}$ for binding and frequently appear to bind to the same conformational states of metaloproteins. Competition between CO and oxygen for oxyreductases is more complex than is the case for liganding to carrier proteins, because the kinetics of substrate oxidation and oxygen reduction affect the apparent affinity of CO relative to apparent affinity for oxygen (Coburn and Foreman 1987).

## CARBON MONOXIDE BODY STORES

There is a normal body store of CO which is derived mainly from two sources: endogenous production of CO and exogenous uptake. Most of the CO body deposits are found in the blood chemically bound to hemoglobin. However, 10-15% of the total body content is located in the extracellar space, probably in combination with myoglobin, cytochrome, catalase, peroxidases, and other heme-bindings; less than 1% is physically dissolved in body fluids (Coburn 1970, Luonmäki 1966, Wennesland *et al.* 1957). This CO distribution is not altered when $PaO_2$ changes within a certain range.

**Endogenous production of CO.** Endogenous production of CO is primarily due to breakdown of the alpha methane bridge of the porphyrin nucleus of hemoglobin in the natural decay cycle of the red blood corpuscles, produced in humans at the rate of 0.42 ml/hr (Coburn 1970). One mole of CO is produced during catabolism of one mole of heme (see Figure 2-1) and provides the major portion of CO that is found in the carboxyhemoglobin of non-smokers, whose hemoglobin levels range from 0.5% to 1.0% (Stewart *et al.* 1974).

Endogenous production of CO was discovered by Sjöstrand *et al.* in 1950 and production rates of endogenous CO in humans are shown in Table 2-2.

TABLE 2-2: ENDOGENOUS PRODUCTION RATES OF CO IN HUMANS (FROM COBURN AND FOREMAN 1987)

| | ml/h | $\mu mol.h^{-1}.kg^{-1}$ |
|---|---|---|
| Normal young men | 0.39-0.44 | 0.28-0.34 |
| Normal young women | | |
| Estrogen phase | 0.32-0.35 | 0.20-0.23 |
| Progesterone phase | 0.62-0.66 | 0.35-0.45 |
| Newborn infants | | 2.5-3.0 |

Degradation of hemoglobin of senescent RBC at the end of their normal life span accounts for approximately 75% of the endogenous CO production. Catabolism of heme-containing compounds within the liver also gives rise to CO. Bone marrow, the third major site of CO production may become important in hemolytic disorders (White 1970). Recent in vitro investigations suggest additional sources, e.g., lipid peroxidation (Wolff and Bidlack 1976).

The level of CO in the blood is a balance between endogenous production and elimination through the lungs as well as oxidation to carbon dioxide (Fig. 2-2). COHb levels due to endogenous production may reach 5% in patients with hemolytic anemia and after microsomal enzyme induction.

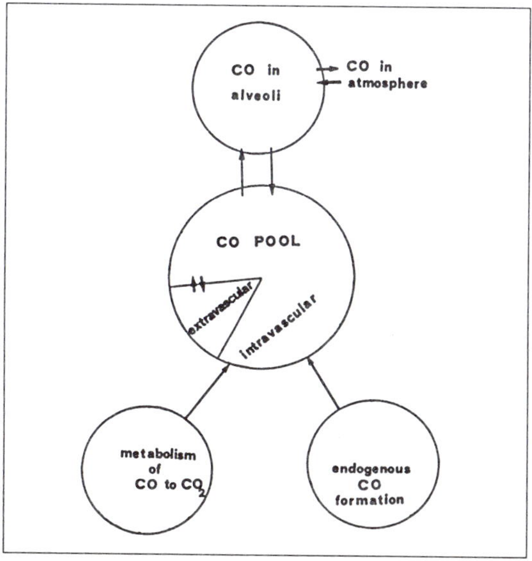


Fig. 2-2: The fate of carbon monoxide (CO) in the human body.

Methyl chloride and other halomethane solvents used in industry are catabolized forming CO and workers exposed to these have elevated COHb levels (Stewart *et al.* 1972). Isotope studies indicated that methylene chloride acts as a direct substrate for metabolic formation of CO (Rodkey and Collison 1977).

**Carbon monoxide production by bacteria.** Some bacteria such as S. fecalis can produce CO coupled with non-enzymatic destruction of heme compounds. This mechanism has been demonstrated in vitro. High levels of COHb have been found at postmortem examination of thoracic cavity fluids of persons who drowned (Shigezane 1986). CO-forming bacteria are listed in Table 2-3. Some bacteria also metabolize CO but this role is insignificant in vivo.

TABLE 2-3
CARBON MONOXIDE (CO) PRODUCING BACTERIA VS NON-CO PRODUCING BACTERIA

| *CO-producing bacteria* | *Non-CO producing bacteria* |
|---|---|
| Streptococcus fecalis (a-hemolytic) | Streptococcus mitis (non-hemolytic) |
| Proteus vulgars | Staphylococcus aureus |
| Proteus mirabilis | Pseudomonas aeruginosa |
| Proteus morganii | Escherichia coli |

## CO UPTAKE AND DISTRIBUTION IN THE BODY

Inhaled CO readily diffuses across the alveolar capillary membrane and resembles oxygen in this respect. Various physiological variables which affect CO uptake and distribution are shown in Table 2-4. Several formulae have been proposed to calculate blood COHb levels from the CO concentrations in the

TABLE 2-4
PHYSIOLOGICAL VARIABLES AFFECTING CO UPTAKE AND DISTRIBUTION

1. CO concentration in the breathed gas mixture and its relation to the partial pressures of oxygen, carbon dioxide, and nitrogen.
2. Density of gas mixture breathed.
3. Temperature and humidity of the air breathed
4. Alveolar ventilation
5. Aleveolar-pulmonary gradient for CO
6. Cardiac output
7. Pulmonary diffusing capacity for CO
8. Speed of reaction of CO with hemoglobin
9. Quantity and speed of flow of blood in the lung capillaries.
10. Hemoglobin and hematocrit values
11. Rate of endogenous CO production
12. Metabolic CO consumption
13. Rate of elimination of CO

inhaled air. Coburn *et al.* (1965) have developed an equation (Coburn-Foster-Kane equation) which takes into consideration pulmonary uptake and excretion, endogenous CO production and dilution in the body CO stores. The basic form is

$$\frac{A[HbCO]_t - B\dot{V}_{CO} - PI_{CO}}{A[HbCO]_0 - B\dot{V}_{CO} - PI_{CO}} = \exp(-t\ A/VbB)$$

where

$$A = \bar{P}c_{O_2}/M\ [HbO_2]$$
$$B = 1/DL_{CO} + PL/\dot{V}A$$

Symbols in the above equations are defined here as follows:

| | |
|---|---|
| $DL_{CO}$ | diffusivity of the lungs for CO, milliliters per minute per mmHg |
| exp | 2.7182, the base of natural logarithms raised to the power of the bracketed expression |
| $[HbCO]_0$ | milliliters of CO per milliliter of blood at beginning of exposure interval |
| $[HbCO]_t$ | milliliters of CO per milliliter of blood at time *t* |
| $[HbO_2]$ | oxyhemoglobin concentration, milliliters of $O_2$ per milliliter of blood |
| *M* | equilibrium constant for reaction of CO with $HbO_2$ |
| $\bar{P}c_{O_2}$ | average partial pressure of $O_2$ in lung capillaries, mmHg |
| $PI_{CO}$ | partial pressure of CO in inhaled air, mmHg |
| PL | barometric pressure minus vapor pressure of water at body temperature, mmHg |
| *t* | exposure duration, minutes |
| $\dot{V}A$ | alveolar ventilation rate, milliliters per minute |
| Vb | blood volume |
| $\dot{V}_{CO}$ | rate of endogenous CO production, milliliters per minute |

This equation has been found to be useful in studying the relative effects of various factors that determine COHb. Effects of different variables on CO uptake are shown in Figure 2-3. It is unlikely that a steady state ever exists in the body where inspired CO is in equilibrium with exogenous COHb in pulmonary capillary blood. There are diurnal variations in COHb levels likely associated with variations in alveolar ventilation and level of physical activity. A person exposed to high concentrations of CO may build up body stores of CO during the day but COHb levels may decrease during the night. COHb levels are raised in smokers (see Chapter 4). Various factors which influence COHb levels in non-smokers are listed in Table 2-5.

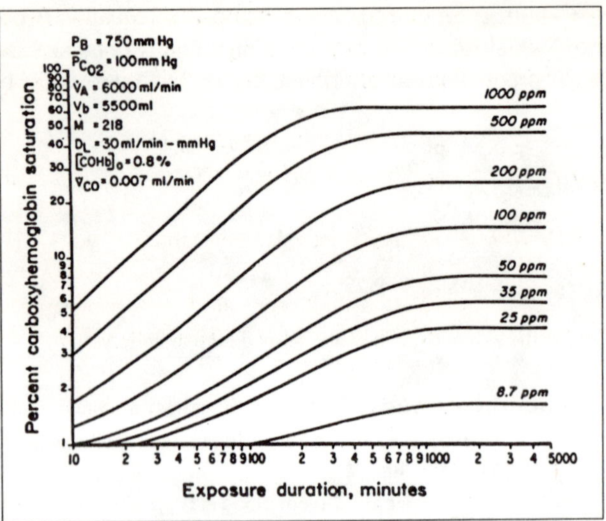


Fig. 2-3: Carbon monoxide uptake a normal human subject as a function of inspired [CO]. PB, barometric pressure; $\bar{P}CO_2$, average partial pressure of $O_2$ in lung capillaries; $\dot{V}A$, alveolar ventilation rate; Vb, blood volume; *M*, equilibrium constant; DL, diffusing capacity of the lungs; $[COHb]_0$, control value prior to CO exposure; $\dot{V}_{CO}$, rate of endogenous CO production. Calculated with the Coburn-Forster-Kane equation. [From Peterson and Stewart 1975]

TABLE 2-5
FACTORS WHICH INFLUENCE COHB LEVELS IN NON-SMOKERS
(BASED ON THE FINDINGS OF KAHN ET AL 1975)

| | | |
|---|---|---|
| 1. Place of work | Industrial workers | 1.38+ 0.04% |
| | Other professions | 0.78+ 0.1% |
| | Unemployed | 0.63+ 0.02% |
| 2. Residence | Urban dwellers have higher COHb levels than rural dwellers | |
| 3. Season | Winter | 1.25% |
| | Spring | 0.58% |
| | Summer | 0.86% |
| | Fall | 0.73% |
| 4. Sex | Women | 0.63% |
| | Men | 0.87% |
| 5. Age | Lowest values in the age group 10-20 yr | |
| 6. Socioeconomic | Middle class has higher COHb than lower or upper classes. | |

Hauck and Neuberger (1984) revised the model of Coburn in order to predict the influence of physiological and breathing parameters on the dynamics of CO intake. Their model seems to describe quite satisfactorily the effect of CO uptake into the blood up to 20% saturation. The manner of CO combination with Hb differs appreciably from that of oxygen at high levels of CO saturation but is virtually the same at low levels of CO saturation (Yamagachi *et al.* 1988).

## CO BINDING TO HEMOGLOBIN

The toxic effects of increased body stores of CO are usually considered to be due to alteration in Hb chemistry. CO is less soluble than oxygen in blood but it combines rapidly with Hb in blood; the time constant at 37°C is less than a few milliseconds (Holland 1962).

The structure of hemoglobin is shown in Figure 2-4. Hemoglobin has been described as a "lung in molecular form" (Perutz and Weiz 1947). It is a complex protein consisting of four polypeptides chains with four heme molecules, each of which is a ferrous complex combined with polypeptide or globulin. The fifth bonding position of the iron atom normally combines with the histidine terminal of the globulin molecule, while the sixth position is available to bond with oxygen. Structures of COHb and $O_2$Hb arc shown in Figure 2-5.

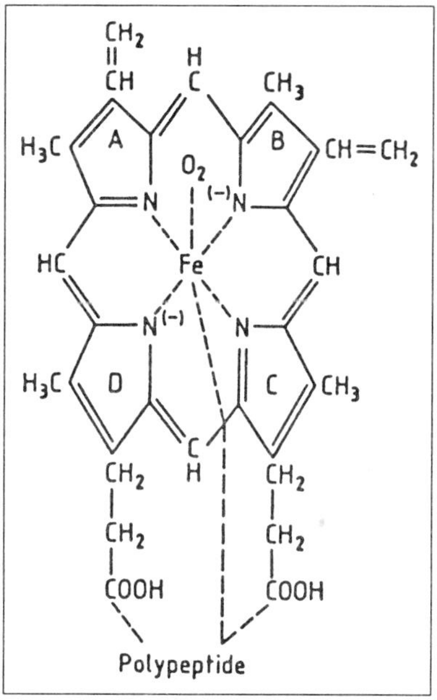

Fig. 2-4: Structure of hemoglobin.

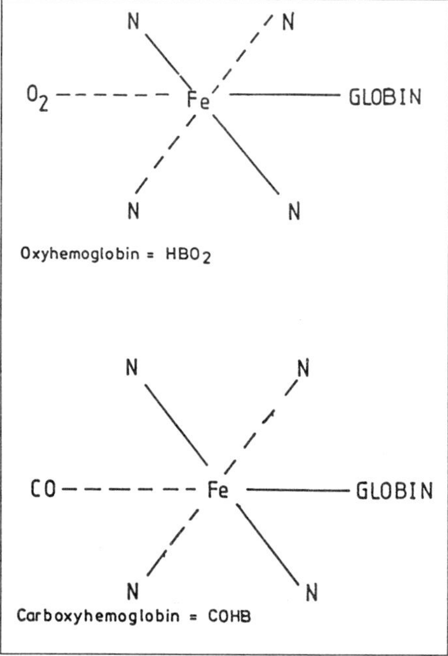


Fig. 2-5: Structures of oxyhemoglobin and carboxyhemoglobin.

An isolated heme solution binds CO about 25,000 times as strongly as oxygen but for myoglobin and hemoglobin the binding affinity of CO is about one-hundredth of this. The structural basis of this is the bent configuration of COHb which weakens the interaction of CO with heme whereas it optimizes the binding of oxygen (Styrer 1981). The iron in the center of the heme is tied to the protein chain on one side.

The other side is bound as $O_2Hb$ in the arterial form (low spin status). In the venous form of Hb, desoxyhemoglobin, the central iron atom has few bindings (high spin status). The reactions of CO with Hb are:

Carbon monoxide + desoxyhemoglobin ⇔ carboxyhemoglobin
Carbon monoxide + oxyhemoglobin ⇔ oxygen + carboxyhemoglobin

The speed with which CO binds to Hb is greater in membrane free Hb suspension as compared to RBC suspension (Forster *et al.* 1957, Holland 1970, Kutchai 1975, Roughton *et al.* 1957). The distribution of COHb in RBCs can be followed by cytological markers (Blackmore 1970). After the inhalation of air containing CO, a few RBCs are "colored" with CO. These colored cells are mostly on the periphery. With further CO intake, the number of colored cells and the intensity of the color increases. From this investigation it follows that the hypoxia of individual cells is not unequivocally correlated with CO saturation of the hemoglobin.

The combination of hemoglobin with CO is governed by Haldane's law (Douglas and Haldane 1912). Accordingly, when a solution containing Hb is saturated with a gas mixture containing oxygen and CO , the relative proportions of the Hb which enter into combination with the two gases are proportional to the relative partial pressures of the two gases (Fig. 2-6), allowing for the fact that the affinity of the CO for Hb is 200 times greater than that of $O_2$. This is expressed by the equation:

$$\frac{COHb}{O_2Hb} = M \frac{pCO}{pO_2}$$

where M, the Haldane constant is 200-240 for human blood (Shepherd 1983)

The rate of formation of COMb can also be expressed by the Haldane equation except that the estimated value of the constant K is 40. Apparently Mb is involved in the oxygen transport mechanism and is ready to deliver oxygen when needed. Examination of $O_2Hb$ and $O_2Mb$ dissociation curves reveals that at $pO_2$ less than 60 mm Hg, $O_2$ has greater affinity for Mb than for Hb.

Combination of CO with Hb has been alleged to lead to anoxemia due to displacement of oxygen from the Hb molecule. However, anoxemia does not appear to be the entire explanation of CO poisoning. A problem posed by Haldane and Priestley in 1935 was: "If the action of carbon monoxide were simply to diminish the oxygen carrying capacity of Hb without other modification of its

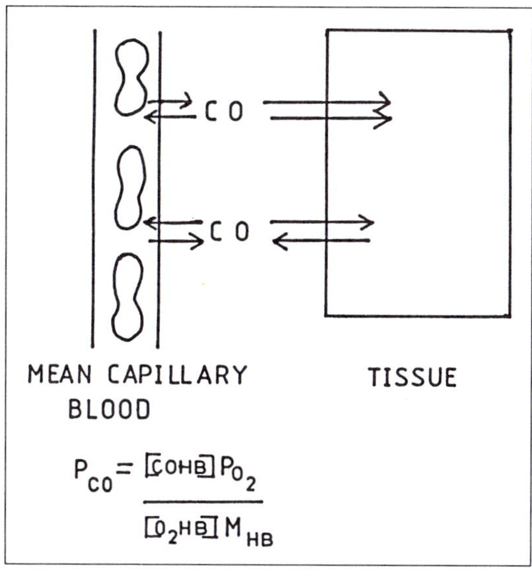


Fig. 2-6: Relative partial pressures of oxygen and carbon monoxide.

properties, the symptoms of carbon monoxide poisoning would be difficult to understand. A person whose blood is half saturated with CO is practically helpless but a person whose Hb percentage is simply diminished to a half by anemia may be going about his work as usual. The key to this seeming paradox is: "CO bound to one heme alters the oxygen affinity of the other heme in the same Hb molecule. Specifically, CO increases the oxygen affinity of the Hb and thereby decreases the amount of oxygen released in actively metabolizing tissues. CO stabilizes the quaternary structure characteristic of $O_2$Hb. In other words, CO mimics oxygen as an allosteric effector" (Styrer 1981).

The primary explanation of this discrepancy is a CO-induced change in the oxygen dissociation curve which shifts to the left (Fig. 2-7). With 50% saturation of the blood with COHb, the oxygen pressure must fall to less than half the usual values in order to dissociate half of the oxygen present. Tissue anoxia is thus far greater than would result from the loss of oxygen-carrying capacity alone. A concentration of 0.06% in the air breathed is enough to block half of the hemoglobin available for oxygen transport.

Important factors which influence the accumulation of COHb are: pH, $pCO_2$, temperature and 2,3 -DPG (diphosphoglycerate). The affinity of oxygen for the hemoglobin is strongly influenced by 2,3-DPG which is located inside the red blood cells. When 2,3-DPG levels rise, as for example, in anaerobic glycolysis, hypoxia, anemia and at high altitudes, affinity of the oxygen for hemoglobin is reduced.

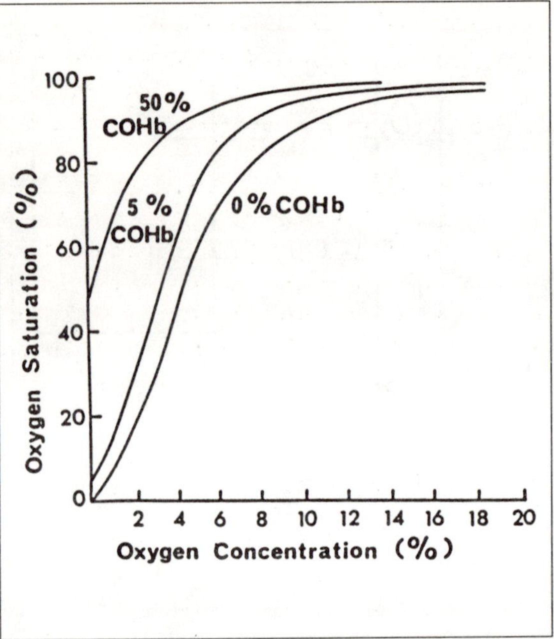


Fig. 2-7: Shift of the oxygen dissociation curve to the left by COHb.

The most obvious physiological effect of CO in vertebrate animals is related to its propensity for competing with oxygen as a ligand in various heme-containing respiratory pigments thus interfering with oxygen transport and utilization. Animals lacking hemoglobin (such as cockroaches) are not known to be affected by CO and animals with great anaerobic capacities such as turtles can tolerate CO atmosphere for several hours (Otis 1970).

## CO ELIMINATION

CO is eliminated mostly by the alveolar capillary membrane. A small amount is oxidized to C02. Fenn (1932, 1970) described the burning of CO in body tissues as shown in the equation

$$2\,CO + O_2 = 2\,CO_2$$

This is a first order reaction and increases as a function of body CO stores. The rate of metabolic consumption of CO is insignificant in the humans. Even with COHb levels as high as 20%, the metabolic rate is small compared with $\dot{V}_{CO}$. Metabolic consumption of CO occurs in the skeletal muscle (Fenn 1932), probably in the mitochondria (Clark 1950). Oxidation of CO to $CO_2$ requires oxygen. Reduced cytochrome will catalyse the oxidation of CO to $CO_2$ after addition of oxygen (Tzagoloff and Whorton 1965). Oxidized heart cytochrome oxidase promotes oxidation of CO to $CO_2$(Young *et al.* 1979). Oxidation of CO to $CO_2$ occurred at CO/$CO_2$ ratios as low as 0.1. The mechanism of CO oxidation is still unclear. It may involve splitting of oxygen atoms by cytochrome-c oxidase similar to that in the normal catalytic reaction but with CO rather than a proton acting as an acceptor of one of the oxygen atoms.

The physiological variables which influence the uptake of CO also affect its elimination. The rate of excretion by the lungs is influenced by the blood levels of COHb, alveolar ventilation, diffusing capacity of the lungs, pulmonary capillary oxygen partial pressure , and other variables. Breathing of 100% oxygen for 3 1/2 hr by a non-smoker reduces the COHb concentration from 0.65 to 0.38% (Stewart *et al.*1973). Elimination half-times of CO are shown in Table 2-6.

TABLE 2-6
HALF-LIFE OF COHB

| | Pressure | Time |
|---|---|---|
| Air | 1 ATA | 5 hr 20 min |
| 100% oxygen | 1 ATA | 1 hr 20 min |
| 100% oxygen | 2 ATA | 23 min |

No definite information is yet available about the release of CO from cytochrome $a,a_3$ in vivo. The half-time for the release of CO from cytochrome $a,a_3$ is about 2 min in vitro (Chance 1970). However, in case of CO poisoning, the rate is much slower.

## RELATION OF CO TO OXYGEN TRANSPORT AND METABOLISM

CO has the following effects on oxygenation in the living organisms:

1. It inhibits the oxygen transport.
2. It inhibits the oxygen availability
3. It inhibits the utilization of oxygen.

Oxygen transport has been discussed in detail elsewhere by Jain (1989). During transit from the arterial to the venous side, the oxygen content of the blood decreases from 20 ml to 15 ml/100 ml. The amount of oxygen extracted is determined by tissue oxygen requirements and blood flow. The oxygen content of arterial blood decreases in proportion to the amount of COHb which carries no oxygen. This produces effects similar to anemia but does not have any of the advantages of improved capillary blood flow resulting from reduced hematocrit. Venous $PO_2$ which is about 40 mm Hg at 75% saturation falls due to a decrease in $PaO_2$ produced by COHb. Since CO differs little from oxygen in molecular weight and diffusing capacity, the two gases arrive in the alveoli and the walls of RBC with their relative proportions changed little from outside air (Forbes 1970).

Once CO has come into contact with Hb, COHb is formed and the effect of this on oxygen transport has already been discussed in the preceding section. The next section will deal with the effect of hypoxia in relation to CO. Following this the cellular effects of CO, i.e., Inhibition of oxygen utilization by the cell. This is considered to be due to binding of CO to cytochrome-c oxidase and cytochrome P-450.

## CO-INDUCED HYPOXIA

Hypoxia has been discussed in detail elsewhere (Jain 1989). Studies with labelled $^{14}C$ in experimental animals have shown that distribution of CO between blood and the tissues remains constant with changes in arterial oxygen tension from 40-650 mm Hg, arterial $CO_2$ from 28-65 mm Hg, and COHb from 0.8 to 55% saturation (Luomanmäki and Coburn 1969). With acute progressive hypoxemia, influx of CO into the myocardium has been shown to occur when oxygen tension falls to less than 30-35 mm Hg (Coburn *et al.* 1973). CO poisoning is primarily a hypoxic lesion caused by replacement of oxyhemoglobin by carboxyhemoglobin (Halebian *et al.* 1986). These authors compared the effects of anoxia induced by 0.5% CO ventilation with that induced by breathing low oxygen mixtures in dogs. They found no significant differences in oxygen consumption or oxygen extraction in the two sets of animals who were subjected to equal reduction of arterial oxyhemoglobin despite the fact that the mode of desaturation was CO poisoning in one group.

One of the problems in studying this subject is that the exact $pO_2$ values inside the living human cell have not been determined as yet. In the heart and the skeletal muscle, $pO_2$ has been estimated to be 10 mm Hg at the sarcolemmal surface and 1 mm Hg in the neighbourhood of the mitochondria. Mb may enhance oxygen flux to the mitochondria by contributing an additional flux of oxyhemoglobin or by acting as a short period oxygen store (Wittenberg *et al.* 1975). In vitro experiments at 25°C have shown that at a critical $pO_2$ between 1-3 mm Hg, CO binds more to Mb than to Hb. This partial blocking of Mb as an intracellular carrier is an indirect explanation of the deleterious effects of CO on cardiac muscle (Agostini *et al.* 1988).

Response to CO hypoxia has also been compared to hypoxic hypoxia. Coburn and Foreman (1987) have reviewed the available evidence on this subject and concluded that cardiovascular autoregulatory mechanisms of adaptation to hypoxia are intact during CO-hypoxia.

Several authors have studied the effects of elevated COHb levels on oxygenation of tissues by measuring tissue $pO_2$. Zorn (1972) showed that brain tissue $pO_2$ decreased linearly with increase in COHb, resulting in 1.8 mm Hg fall per 1% saturation increase in COHb. This suggests that a major mechanism for CO toxicity is the effect on capillary $pO_2$ with resultant fall in tissue $pO_2$.

In addition to acute compensatory mechanisms for CO-induced hypoxia, animals chronically exposed to CO have additional mechanisms that would allow them to be exposed to CO concentrations that would kill them if they had not been exposed to CO previously. Mice adapted to high altitude have been shown to be resistant to CO poisoning (Clark and Otis 1952). These mechanisms include increase in Hb and RBC count, increase in cytochrome-c concentration, and an increase in myocardial capillary density.

2,3-diphosphoglycerate (DPG) levels are generally increased in chronic hypoxic hypoxia. The hypoxia-driven enhancement of DPG synthesis is not dependent upon an increase in reduced Hb. It appears that there are two drives governing DPG concentrations at higher COHb levels; decrease in available binding sites and hypoxia-induced DPG synthesis. The former suppresses the DPG synthesis more potently than the latter and thus there is no significant change in DPG (Dinman *et al.* 1970).

There are few experimental studies where metabolic effects of CO have been studied in tissues. A few studies done on the brain have utilized high $P_{CO}$ levels with resulting fall in tissue $pO_2$ and showed no difference between hypoxic hypoxia and CO-hypoxia (Winston-Roberts 1978). It is likely that the effects in these studies were dominated by tissue hypoxia or anoxia. Investigations by Piatadosi *et al.* (1988a) suggest that CO alters brain metabolism in vivo independently of the COHb-related decrease in oxygen delivery.

Cerebral hypoxia resulting from CO differs from the hypoxic or the anemic hypoxia in that the cerebral oxygen delivery is increased but the cerebral $O_2$ extraction fraction is decreased. Fetal sheep hemoglobin has a $P_{50}$ ($pO_2$ at 50% saturation of non-CO bound sites) value of approximately 14 torr while adult human hemoglobin has a $P_{50}$ of 35-44 torr. In exchange transfusions on newborn lambs Koehler *et al.* (1983) demonstrated that the effect of $O_2$Hb dissociation curve after the transfusion and a leftward shift after CO are similar in regard to cerebral oxygen delivery and fractional oxygen extraction. These findings support the presence of a highly sensitive tissue oxygen-dependent mechanism regulating the cerebral circulation. The fetal condition of low arterial oxygen content and an oxygen dissociation curve which is shifted to the left may have provided the newborn lamb with a microcirculation better suited for maintaining CMRO during CO hypoxia (Koehler *et al.* 1984).

In spite of an adequate cerebrovascular oxygen tension, oxygen delivery may be insufficient to meet the needs of specific brain structures. Furthermore, binding of CO to intracellular cytochrome oxidase may occur more readily in the brain under conditions of extreme hypoxia for the following reasons:

1. Cytochrome oxidase has a Michaelis-Menton constant ($K_m$) for $O_2$ below 1 mm Hg in vitro (Chance 1955).

2. Only the reduced cytochrome-a, iron porphyrin center found in hypoxic state binds CO in vitro (Yoshikawa and Caughey 1982).

3. Cytochrome oxidase has a Warburg partition coefficient K (a measure of the CO affinity of an intracellular compound and is approximately equal to a reciprocal of K) in the range of 2-28 (Wohlraband Ogunmula 1971).

A K value of 2 for cytochrome oxidase means that the mitochondrial $P_{CO}$ would have to be at least two times the mitochondrial $pO_2$ to bind 50% of oxidase. This occurs in extreme hypoxia only.

The loss of $O_2$ utilization in CO-blocked oxidase molecule is offset by an increase in $O_2$ uptake by adjacent oxidase resulting in no net change in CMRO. Animal studies have shown that CMRO does not decline till COHb concentrations near 50% (Pitt *et al.* 1979). This percentage corresponds to a maximum of about 50% cytochrome $a,a_3$ -CO. According to Piantadosi (1987), the ability of a cell to avoid bioenergetic failure depends, not only on the availability of $O_2$ but to some extent on the concentrations of CO and cytochrome oxidase.

## CELLULAR EFFECTS OF CO

Some effects effects of CO are not explained by COHb-related decrease in tissue oxygen tensions, particularly, the effects of COHb at concentrations below 5%. The clinical manifestations of CO-toxicity also do not correlate with COHb levels (Sokal 1975). Late sequelae of CO poisoning are known to occur after COHb levels have declined. Warburg (1927,1926) demonstrated that CO competes with oxygen for the reduced form of cytochrome-$a_3$ oxidase which is the terminal enzyme of the cellular respiratory chain. It is also known that CO binds to reduced cytochrome-c oxidase (Keilin and Hartree 1939) and reduced cytochromes of P-450 type (Omura *et al.* 1965). Cytochrome prefers oxygen to CO by 9:1 and this may explain the disparity between COHb levels and clinical effects of CO poisoning. The possibility that CO inhibits mitochondrial electron transport in vivo is interesting because of the close relationship between the respiratory chain function and the cellular energy metabolism (Fig. 2-8). Carbon monoxide combines with cytochrome-$a_3$ oxidase and cytochrome P-450, thus blocking cellular oxidation and causing cellular hypoxia. These basic mechanisms have been confirmed by Chance (1970). Organs with high metabolic rate such as the heart and the brain, are particularly affected by CO (Myers *et al.* 1975). In vivo, CO mediated cytochrome-b reduction reduction responses in the rat brain have been shown to be blocked by pretreatment with cytochrome-b inhibitor, antimyecin A (Piantadosi *et al.* 1987). Piantadosi *et al.* (1988b) have postulated that cyanide resistant CO binding component might account for the in vivo cytochrome $a,a_3$ interaction and directly or indirectly modulates cytochrome to reduction responses to CO exposure.

Molecular structure and orientations of the cytochrome-$a_3$ oxidase account for the affinity between the two under physiological conditions (Argade *et al.* 1984). Changing the valence state of the cytochrome-a has a significant effect on its CO-binding properties (Clore and Chance 1980). CO binding to cytochrome P-450

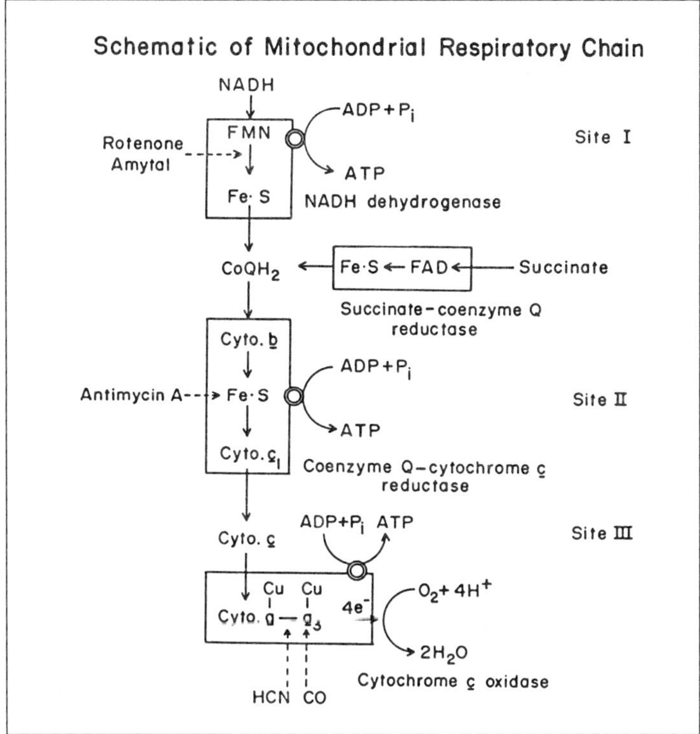


Fig. 2-8: Sites of action of CO on the mitochondrial respiratory chain function. *(Piantadosi 1987, by permission).*

enhances suppression of hexobarbital oxidation due to hypoxia (Takano *et al.* 1985). Oertle *et al.* (1985) have studied the kinetics of CO binding to cytochrome P-$450_{PB}$ and its enzymatically inactive form P-$420_{PB}$ and concluded that it was a simple bimolecular process. Optical spectroscopy studies performed by the authors showed the existence of an equilibrium between the two forms of P-450.

Chronic CO-induced hypoxia induces biochemical adaptations including an increase in both Hb and Mb. However, cytochrome c oxidase activity remains depressed indicating that it does not play a role in the adaptive response (Pankow and Pansold 1984). Cytochrome P-450 increases in the livers of animals exposed to certain drugs, chemicals, and environmental pollutants. This is considered to be an adaptive response. Such a response is not seen after repeated CO intoxications in experimental animals (Pankow *et al.* 1982).

The exact mechanisms responsible for the in vivo mitochondrial reactions to CO and their costs to the cell are far from clear. Piantadosi *et al.* (1985 have studied the intracellular effects of CO in a fluocarbon-circulated rat preparation. Reflectance spectrophotometry has shown that mitochondrial cytochrome redox response occurs at CO:$O_2$ ratios that translate to COHb levels of 30%. This happens during fluocarbon-circulation at high venous $pO_2$; therefore, when $O_2$ transport depends on hemoglobin and COHb formation formation causes tissue $pO_2$ to fall, the observed CO-cytochrome effects should appear at COHb levels well below 30% (Piantadosi 1987).

## INTERACTION OF CO WITH OTHER TOXIC COMPOUNDS

Interaction between CO and $CO_2$ has been studied because of the simultaneous exposure to these gases in smoke inhalation injury. $CO_2$ is a normal byproduct of cellular metabolism and is extremely important in the regulation of vital functions of the body. In experimental studies, it was shown that addition of non-lethal concentrations $CO_2$ (1.7 to 17.3%) to sublethal concentrations of CO (2500 to 4000 ppm) caused death of exposed rats, suggesting a synergistic action of the two gases.

Studies on rats have shown that combination of ethanol (alcohol) with CO produces additive effects with respect of increase of indicator enzymes if at least 50% of the Hb is saturated with CO and if the noxious agent is administered in large doses (Pankow *et al.* 1974).

COHb concentrations increase after inhalation exposure to dichlormethane ($CH_2Cl_2$). The biotransformation of dichlormethane to CO is mediated by hepatic cytochrome P-450 dependent enzyme systems. Fodor *et al.* (1973) tested the combined effect of CO and dichlormethane in rats. After exposure to 116 mg CO + 3.5 gm $CH_2Cl_2/m^3$ of breathed air, the COHb rose to 20% (instead of 10% as expected from CO alone). Simultaneous administration of isoniazid (which induces an isoenzyme of P-450) reduces the dichlormethane metabolism as shown by lower COHb levels. This is considered to be possibly due to non-specific inhibition of dichlormethane conversion to CO in the liver (Pankow 1988).

Cyanide also poisons cytochromes. There is an interaction between cyanide and CO poisonings as seen clinically in fire victims. CO and cyanide are physiologically additive in increasing CBF and decreasing cerebral oxidative metabolism (Pitt *et al.* 1979). COHb levels do not accurately predict cyanide exposure even though both are common constituents of smoke. If cyanide toxicity is suspected, direct assay for cyanide is recommended (Barillo *et al.* 1986).

## MEDICAL USES OF CO

The main medical use of CO is for determination of lung diffusion capacity. This test is based on Ficks equation which describes the diffusion of gases through membranes of given substances $\frac{dn}{dt}$

$$\frac{dn}{dt} = -q \, . \, D \, . \, \frac{dc}{dx}$$

where q is area

D is diffusion coefficient

$\frac{dc}{dx}$ is concentration gradient

Inhaled CO is bound to the Hb completely. When the alveolar partial pressure of CO ($pA_{CO}$) is not high, the diffusion capacity of the lung ($D_LCO$) is:

$$D_L\,CO = \frac{CO\text{ - Intake (ml/min)}}{pA_{CO} - pC_{CO})}$$

The capillary CO tension ($pC_{CO}$) is determined from the COHb content of the peripheral blood. The median CO diffusion capacity in the healthy individual is 143-263 $ml/min^{-1}.\,k\,Pa^{-1}$ (19-35 $ml.\,min^{-1}\,.\,mm\,Hg^{-1}$ ).

The original technique was described by Krogh (1914) and several modifications are used now. Uchida *et al.* (1986) have described a rebreathing method for simultaneous measurement of cardiac output and pulmonary diffusion capacity for CO. The transfer factor for CO ($T_L$ CO) rises during exercise in a curvilinear manner with $K_{CO}$ (transfer coefficient for CO) up to $VO_2$ max (Kendrick *et al.* 1986).

In anemic patients, the lowered Hb levels affect the CO diffusion capacity and a correction factor needs to be used (Petermann 1985). In patients with obstructive ventilatory defect, a low CO diffusion capacity (40% or less) is associated with an increased mortality (Peter-Golden *et al.* 1984).

CO has also been used for assessment of total blood cell volume, total Hb, and erythropoietin but better methods are available for these tests.

Under conditions of hypovolemic hypotension where a large fraction of the total body CO store is found in the extravascular tissue, calculations of blood volume from dilution of CO give figures that grossly overestimate the true blood volume (Coburn *et al.* 1971).

## CONCLUDING REMARKS

Considerable information is available in the literature about the mechanism of CO toxicity. Most of the toxic effects of CO are considered to be due to binding of this gas to Hb with COHb formation. The deleterious effects of CO are often referred to as CO hyposia with the implication that pathophysiology of CO poisoning involves inhibition of oxygen transport from the blood to the tissues with resultant fall in the tissue partial pressures of oxygen. Recent evidence points to the role of CO combination with cytochrome oxidase and inhibition of cell utilization of oxygen. Although the half-life of COHb is known but that of CO bound with cytochrome oxidase is not known. More research is needed to determine this as it may be an important factor in the genesis of late sequelae of CO poisoning and may also provide a rational basis for determining the duration of treatment of CO poisoning by hyperbaric oxgenation.

The role of P-450 in CO poisoning needs further clarification. Because of the genetic diversity of this pigment during evolution, there are differences between human and rodent P-450s (Gonzalez 1989). Monkeys may be more suitable experimental animals for studying the response of P-450 to various toxins.

## REFERENCES

Agostoni A, Perrella M, Sabbioneda L et al (1988) CO binding to hemoglobin and myoglobin in equilibrium with a gas phase of low $PO_2$ value. J Appl Physiol 65:2513-2517.

Argade PV, Ching YC, Rousseau DL (1984) Cytochrome $a_3$ structure in carbon monoxide bound cytochrome oxidase. Science 225:329-331.

Barillo DJ, Goode R, Rush BF et al (1986) Lack of correlation between carboxyhemoglobin and cyanide in smoke inhalation injury. Current Surgery 43:421-423.

Bianconi A, Conjiu-Castellano A, Durham PJ et al (1985) The CO bond angle of carboxyhemoglobin determined by angular-resolved XANES spectroscopy. Nature 318:685-687.

Blackmore DJ (1970) Distribution of HbCO in human erythrocytes following inhalation of CO. Nature 227:386.

Caughey WS (1970) Carbon monoxide binding in hemoproteins. Ann NY Acad Sc 174:148-153.

Chance B, Williams GR (1955) Respiratory enzymes in oxidative phosphorylation. IV. The steady state. J Biol Chem 217:409-427.

Chance B, Erecinska M, Wagner M (1970) Mitochondrial Responses to Carbon Monoxide Toxicity. Annals of the New York Academy of Sciences 174: 193-204.

Clark RT (1950) Evidence for the conversion of carbon monoxide to carbon dioxide in the intact animal. Am J Physiol 162:560-564.

Clark RT, Otis AB (1952) Comparative studies on the acclimitization of mice to carbon monoxide and to low oxygen. Am J Physiol 169: 285-294.

Clore GM, Chance EM (1980) CO binding to mitochondrial mixed valence state cytochrome oxidase at low temperatures. Biochem Biophys Acta 590:34-49.

Coburn RF, Forster RE, Kane PB (1965) Considerations of the physiological variables that determine blood carboxyhemoglobin concentration in man.
J Clin Invest 44:1899-1910.

Coburn RF (1970) The carbon monoxide body stores.
Ann NY Acad Sc 174:11-22.

Coburn RF, Wallace HW, Abboud R (1971) Redistribution of body carbon monoxide after hemorrhage.
Am J Physiol 220:868-874.

Coburn RF, Ploegmakers F, Gondrie P et al (1973) Myocardial myoglobin oxygen tension.
Am J Physiol 224: 870-876.

Coburn RF, Forman HJ (1987) Carbon Monoxide Toxicity. In, Farhi LE, Tenney SM (eds): The Respiratory System, Vol IV, section 3 of Handbook of Physiology, Bethesda, Maryland, American Physiological Society, 1987, pp. 439-456.

Dinman BD, Eaton JW, Brewer GJ (1970) The effects of carbon monoxide on DPG concentrations in the erythrocyte.
Ann NY Acad Sc 174:246-251.

Douglas CG, Haldane JS, Haldane JBS (1912) The Laws of Combination of Haemoglobin with Carbon Monoxide and Oxygen.
Journal of Physiology 44: 275-304.

Fenn WO, Cobb DM (1932) The stimulation of muscle respiration by carbon monoxide.
Am J Physiol 102:379-392.

Fenn WO (1970) The burning of CO in tissues.
Ann NY Acad Sc. 174:64-71.

Fodor G. Prajsner D, Schlipköter HW (1973) Endogene CO-Bildung durch inkorporierte Halogenkohlenwasserstoffe der Methanereihe.
Staub Reinhalt Luft 33:258-259.

Forbes WH (1970) Carbon monoxide uptake via the lungs.
Ann NY Acad Sc 174:72-75.

Forster RE, Roughton FJW, Kreuzer F et al (1957) Photocalorimetric determination of rate of uptake of CO and $O_2$ by reduced human red cell suspensions at 37° C.
J Appl Physiol 11:260-268.

Gonzalez FJ (1989) The molecular biology of cytochrome P-450s.
Pharmacol Rev 40:243-288.

Haldane JS, Priestley JG (1935) Respiration, 2nd ed. Oxford, Clarendon Press.

Halebian P, Robinson N, Barie P et al (1986) Whole Body Oxygen Utilization During Acute Carbon Monoxide Poisoning and Isocapneic Nitrogen Hypoxia.
The Journal of Trauma 26: 110-117.

Hauck H, Neuberger M (1984) Carbon monoxide uptake and the resulting carboxyhemoglobin in man.
Eur J Appl Physiol 53: 186-90.

Holland RAB (1970) Reaction rates of carbon monoxide and hemoglobin.
Ann NY Acad Sc 174:154-171.

Holland RAB (1969) Rate of $O_2$ dissociation from $O_2Hb$ and relative combination rate of CO and $O_2$ in mammals at 37°C.
Resp Physiol 7:30-42.

Jain KK (1989) Oxygen in Physiology and Medicine, Springfield, Illinois, Charles C. Thomas.

Kahn A, Rutledge RB, Davis GL et al (1975) A study of carbon monoxide sources in St. Louis Metropolitan population and some policy implications.
CUERS REport N. 4:1-105.

Kendrick AH, Cullen J, Green H et al (1986) Measurement of single breath carbon monoxide transfer factor (diffusing capacity) during progressive exercise.
Bull Eur Physiopath Resp 22:365-370.

Keilin D, Hartree EF (1939) Cytochrome and cytochrome oxidase.
Proc Roy Soc Lond Ser B 127:167-191.

Koehler RC, Traystman RJ, Rosenberg AA et al (1983) Role of O2-hemoglobin affinity on cerebrovascular response to carbon monoxide hyposia.
Am J Phsyiol 245:H1019-H1023.

Koehler RC, Traystman RJ, Zeger S et al (1984) Comparison of cerebrovascular response to hypoxic and carbon monoxide in newborn and adult sheep.
J. Cereb Bloodflow Met 4:115-122.

Krog M (1914) The diffusion of gases through the lungs of man.
J Physiol (Lond) 49:271-300.

Kutchai H (1975) Role of red cell membrane in oxygen uptake.
Resp Physiol 23:121-132

Levin BC, Paabo M, Joshua MS et al (1987) Toxicological interactions between carbon monoxide and carbon dioxide.
Toxicology 47:135-164.

Luomanmäki K, Coburn RF (1969) Effects of metabolism and distribution of carbon monoxide on blood and body stores.
Am J Physiol 217:354-363.

Luomanmäki K (1966) Studies on the metabolism of carbon monoxide.
Ann Med Exper et Biol Fenn 44 (Suppl 2): 1-55.

Oertle M, Richter C, Winterhalter KH et al (1985) Kinetics of carbon monoxide binding to phenobarbital-induced cytochrome P-450 from rate liver microsomes.
Proc Nat Acad Sci USA 82:4900-4904.

Oh SH, Fisher GB, Carpenter JE et al (1986) Comparative kinetic studies of $CO-O_2$ and CO-NO reactions over single crystal and supported rhodium catalysts.
J Catalysis 100:360.

Omura T, Sata R, Cooper DY et al (1965) Function of cytochrome P-450 of microsomes.
Fed Proc 24:1181-1189.

Otis AB (1970) The physiology of carbon monoxide poisoning and evidence of acclimatization.
Ann NY Acad Sc 174: 242-245.

Pankow D, Ponsold W, Fritz H (1974) Combined effects of carbon monoxide and ethanol on the activities of leucine aminopeptides and glutamic-pyruvic tansaminase in the plasma of rats.
Arch Toxicol 32:331-340.

Pankow D, Schiller F, Müller D (1982) Effect of repeated carbon monoxide exposure to rates on cytochrome P-450 concentration and activities of mono-oxygenases in the liver.
Acta Biol Med Germ 41:935-939.

Pankow D (1988) Enhancement of dichlormethan-induced carboxyhemoglobinemia by isoniazid pretreatment.
Biomed Biochim Acta 47:293-295.

Pankow D, Ponsold W (1984) Effect of carbon monoxide exposure on heart cytochrome c oxidase activity of rats.
Biomed Biochim Acta 43: 1185-1189.

Paul J, Smith ML, Paul KG (1985) The vibrational bands of carbon monoxide bound to hemes or metal surfaces.
Biochimica et Biophysica Acta 832:257-264.

Peterson JE, Steward RD (1975) Predicting the carboxyhemoglobin levels resulting from carbon monoxide exposures.
J Appl Physiol 39:633-638.

Petermann W (1985) Effect of low hemoglobin levels on the diffusing capacity of the lungs for CO
Respiration 47:30-38.

Peters-Golden M, Wise RA, Hochberg MC et al (1984) Carbon monoxide diffusing capacity as a predictor of outcome in systemic sclerosis.
Am J Med 77:1027-1034.

Perutz MF, Weiz O (1947) Crystal structure of carboxyhemoglobin.
Nature 169:786-787.

Piantadosi CA, Sylvia AL, Saltzman HA et al (1985) Carbon monoxide-cytochrome interactions in the brain of the fluocarbon-perfused rat.
J Appl Physiol 58:665-672.

Piantadosi CA, Lee PA, Sylvia AL (1988a) Direct effects of CO on cerebral energy metabolism in bloodless rats.
J Appl Physiol 65:878-887.

Piantadosi CA, Sylvia AL, Jöbsis-Vandervliet F: Differences in brain cytochrome responses to carbon monoxide and cyanide in vitro.
J Appl Physiol 62:1277-1284.

Pitt BR, Radford EP, Gurtner GH et al (1979) Interaction of carbon monoxide and cyanide on cerebral circulation and metabolism.
Arch Environmental health 34:354-359.

Rodkey FL, Collison HA (1977) Effect of dihalogenated methanes on the in vivo production of carbon monoxide and methane by rats.
Toxicol Appl Pharmacol 40:39-47.

Roughton FJW, Forster RE, Cander L (1957) Rate at which carbon monoxide replaces oxygen from combination with human hemoglobin in solution and in the red cell.
J Appl Physiol 11:269-276.

Shepherd RJ (1983) Carbon Monoxide — the silent killer. Springfield, Illinois, Charles C Thomas.

Shigezane J (1986) Postmortem Formation of Carbon Monoxide by Bacteria.
JPN J Legal Med 40: 111-118.

Sjöstrand T (1950) Endogenous formation of carbon monoxide.
Acta Physiol Scand 22:137-141.

Steward RD, Fisher TN, Hosko MH et al (1972) Carboxyhemoglobin elevation after exposure to dichlormethane.
Science 176:295-296.

Steward RD, Baretta ED, Platta LR et al (1974) Carboxyhemoglobin Levels in American Blood Donors.
JAMA 229: 1187-1195.

Styrer L (1981) Biochemistry, 2nd ed. San Francisco, W.H. Freeman & Co. pp 54-55.

Takano T, Motohashi Y, Miyazaki Y (1985) Direct effect of carbon monoxide on hexobarbital metabolism in the isolated perfused liver in the absence of hemoglobin.
J Toxicol Environ Health 15:847-854.

Tzagoloff A, Whorton DC, (1965) Studies on the electron transfer system. LXII. The reaction of cytochrome oxidase with carbon monoxide.
J Biol Chem 240:2628-2633.

Uchida K, Shibuya I, Mochizuki M (1986) Simultaneous measurement of cardiac output and pulmonary diffusing capacity for CO by a rebreathing method.
Jap J Physiol 36:657-670.

Warburg O (1926) Über die Wirkung des Kohlenoxyds auf den Stoffwechsel der Hefe.
Biochem Z 177:471-486.

Warburg O (1927) Über Kohlenoxidwirkung ohne Hämoglobin and einige Eigenschaften des Atmungsfermentes.
Naturwiss 15:546.

Wennesland R, Nomof N, Brown E et al (1957) Distribution of CO and radiochromium in blood and tissues of rabbit and dog. I. carbon monoxide.
Proc Soc Exp Biol Med 96:655-657.

White P (1970) Carbon monoxide production and heme catabolism.
Ann NY acad Sc 174:23-31.

Winston JM, Roberts RJ (1978) Glucose catabolism following carbon monoxide or hypoxic hypoxia exposure.
Bioch Pharmacol 27:377-380.

Wittenberg BA, Wittenber JB, Caldwell PRB (1975) Role of myoglobin in oxygen supply to red skeletal muscle.
J Biol Chem 250:9038-9043.

Wohlrab H, Ogunmola GB (1971) Carbon monoxide binding studies of cytochrome $a_3$, hemes in intact rat liver mitochondria.
Biochemistry 10:1103-1106.

Wolfe DG, Bidlack WR (1976) The formation of carbon monoxide during peroxidation of microsomal lipids.
Biochem Biophys Res Commun 73:850-857.

Yoshikawa S, Caughey WS (2982) Heart cytochrome c oxidase
J Biol Chem 257:412-420.

Young LJ, Choc MG, Caughey WS (1979) Role of oxygen and cytochrome c oxidase in the detoxification of CO by oxidation to $CO_2$. In, Caughey WS (ed)
Biochemical and Clinical aspects of oxygen, New York, Academic Press, pp. 355-361.

Zorn H (1972) The partial oxygen pressure in the brain and liver at subtoxic concentrations of carbon monoxide.
Staub-Reinhalt Luft 32:24-29.

# Chapter 3

## ENVIRONMENTAL CO AND EPIDEMIOLOGY OF CO POISONING

### CARBON MONOXIDE IN THE ENVIRONMENTS

Carbon monoxide has been present in trace concentrations in the atmosphere even before the dawn of human civilization. Next to carbon dioxide, it is the most abundant and widely distributed air pollutant found in the lower atmosphere.

#### Man-made CO Pollution

The major sources of atmospheric pollution by CO are man-made as shown in table 3-1. The total emission from these sources is estimated to be 700-2,000 million tons/year (Graedel and Crutzen 1989). Automobile exhaust accounts for more than half of the global output of CO. The United States account for about one-third of the global output of CO (WHO 1977).

TABLE 3-1
MAN MADE SOURCES OF ATMOSPHERIC CARBON MONOXIDE

1. Incomplete combustion of carbonaceous fuel used for transport e.g. automobile exhaust
2. Industrial plant exhausts
3. Detonation of explosives
4. Smoking of cigarettes
5. Burning of solid waste
6. Defective household heating and cooking systems

#### The Automobile as a Source of Atmospheric CO

Automobile exhaust is the commonest source of atmospheric CO pollution in the industrialized urban areas of the Western countries. It accounts for 90% of the atmospheric content of the atmosphere of a city. In the absence of an emission control device, an American car of the 1960's produces about 0.37 Kg of CO from each litre of gasoline that is consumed. The exhaust contains 2-10% by volume of CO. In contrast, diesel engine exhaust contains 0.1% CO. The CO concentration depends mainly on air to fuel ratio. Improvements in engine design in the seventies led to a substantial increase in the air to fuel ratio and a reduction of 70% in the emissions. The use of catalytic converters (which convert CO to $CO_2$ and water) and emission control testing, which is mandatory in some states, has helped to reduce the atmospheric CO contamination. The intent of the United States Environmental Protection agency is to reduce the average emission of CO to 13 g/Km by the year 1995. In Switzerland, according to the "Schweizerische Luftreinhalt-Verordnung," the automobile exhaust should not contain more than 8 mg/m$^3$ of CO (Wanner *et al.* 1984).

At any location, the concentration of carbon monoxide due to motor vehicle traffic depends on the following specific variables:

a. Number of vehicles operating
b. Engine characteristics of the operating vehicles
c. Speed of traffic and gradient
d. Temperature (as it affects the operating efficiency).

Wanner (1989) has measured the CO exposure in Zurich city (Fig. 3-1). The highest CO concentrations were found in the car interior. The car driver is exposed to a CO concentration which is double that for the pedestrian and 1 1/2 times that for the bicycle rider in the traffic. (Isles *et al.* 1989).

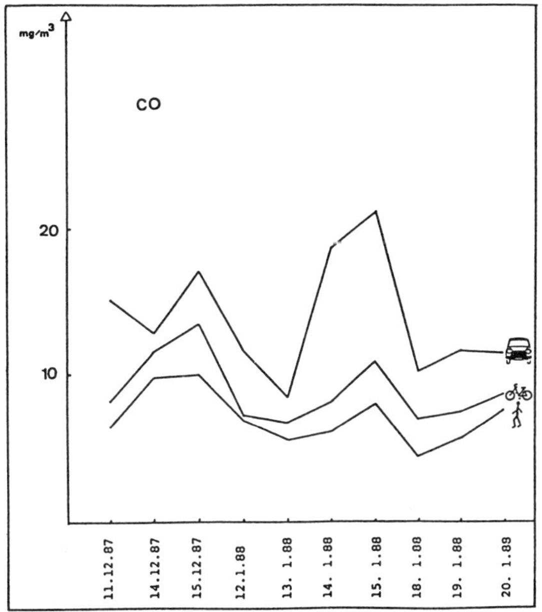


Fig. 3-1: Carbon monoxide exposure measured in Zürich, Switzerland in winter 1987/88. Car driver, bicycle rider and pedestrian. The car driver has the highest exposure to CO.

*Illustration by courtesy of Prof. H. U. Wanner, ETH (Federal Institute of Technology), Zürich.*

On busy city intersections, CO concentrations as high as 0.03% have been observed. A pedestrian on a street with heavy automobile traffic is exposed to CO. A concentration of 20-40 mg/ml can raise COHb 1 1/2 to 2-fold within one hour. Jogging in this environment increases the CO intake and further raises the COHb. Persons doing manual work on streets with heavy automobile traffic can suffer a rise of COHb to toxic levels. Jogging in Central Park, New York City, can be more dangerous than walking or just standing around. Smoking in such environments further aggravates CO intoxication and COHb levels of 13% can be reached. The risk of CO intoxication is also higher in the cities located at high altitude. The atmosphere in Mexico city (altitude 7350 ft) has not only 30% less oxygen as compared to the sea level, but the car engine also emits twice as much at this altitude than at sea level. Breathing in such an atmosphere is equal to smoking two packets of cigarettes daily.

Other critical areas for CO intoxication are the Autobahn (highway) tunnels in Switzerland. The longest of these, the Gotthard tunnel (length 17 km), has traffic jams during the peak holiday season and CO concentrations of 015mg/ml have been recorded in the tunnel. (Greenpeace 1988).

## Natural Sources of CO

The natural sources of CO are only a minor contribution to the atmospheric CO pool. These sources are listed in Table 3-2.

TABLE 3-2
NATURAL SOURCES OF CARBON MONOXIDE

A. Geophysical
   1. Volcanic gases
   2. Marsh gases
   3. Natural gases in coal mines
   4. Photodissociation of CO in upper atmosphere
   5. Formation of CO during electrical storms
   6. Forest fires

B. Biological
   1. From vegetation during seed germination
   2. From marine brown algae or kelp
   3. Marine hydrozoans: e.g. jelly fish
   4. Endogenous CO production by land animals

## Concentration of CO in the Atmosphere

Natural background levels of CO are low and vary from 0.01 to 0.2 ppm (part per million) with an average of 0.1 ppm. Concentrations in urban areas are closely related to motor traffic density. They also vary with time and are dependent on meteorological factors. There are diurnal, weekly and season variations. They are at their highest levels during peak rush hour traffic. CO and other pollutants in the atmospheres of cities rise during prolonged periods of air stagnation whereas wind disperses the pollutants. A gradual decrease in CO concentration occurs with increasing altitude ranging from approximately 0.08 ppm at 4 Km to 0.04 ppm at 15 Km (Goldman *et al.* 1973). High concentrations are noted in the Northern hemisphere where 90% of the world's fuel consumption takes place. The concentration at the equator is about 0.1 ppm and in the Southern hemisphere, 0.04

ppm concentration of CO has been recorded at 50°S latitude (Robinson and Robins 1970).

In the past, high concentrations of CO in the Southern Hemisphere had been considered unlikely because it was less industrialized. The burning of tropical rain forest, however, generates as much CO as burning of fossil fuels does. Measurement of CO by special devices in space (Newell *et al.* 1989) have recorded the highest concentrations of CO in area with little or no industry or automobile traffic. Many of such areas lay in the Southern Hemisphere or the tropics.

Contributions of various sources of CO for the community are shown in Table 3-3 and Fig. 3-2.

TABLE 3-3
POSSIBLE CONTRIBUTIONS OF VARIOUS SOURCES OF CARBON MONOXIDE FOR THE COMMUNITY
*(from Goldsmith 1970, by permission)*

| | *Number* | *Intensity* | *Time Course* |
|---|---|---|---|
| Cigarette smoking | 40-60% of adults | maxima of about 15% COHb | intermittent diurnal patterns<br>duration-decades |
| Occupation | ? 750,000 | usually <10% COHb high concentrations may cause death | 8 hours or less per day, 5 days per week<br>duration-variable |
| Household | ? 100,000<br>no data | usually <2% COHb high concentrations may cause death | occasional, 4-8 hours may be repeated |
| Community air pollution (auto exhaust) | ? 30,000,000<br>no data | usually <2% COHb high concentrations may cause illness | unpredictable (4-8 hours) |

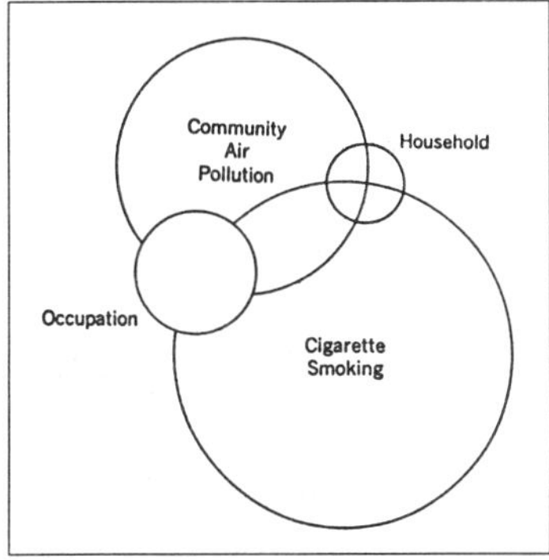


Fig. 3-2: Proportion of various sources of carbon monoxide pollution in the community. See Table 3-3.
*(Reproduced from Goldsmith, 1979, by permission)*

CO concentrations are usually reported in terms of 8 h average concentrations. This average time has been chosen because it is the time required for the COHb levels in the body to reach equilibrium with the ambient CO concentrations.

**Chemical Reactions of CO in the Atmosphere**

CO is oxidized very slowly in the lower atmosphere by two reactions:

1. $CO + O_2 \rightarrow CO_2 + O$
2. In presence of moisture: $CO + H_2O \rightarrow CO_2 + H_2$

These reactions are catalysed by oxides of magnesium and copper. Ozone will oxidize CO to $CO_2$ but the rate is extremely slow at atmospheric temperatures and concentrations. Some reactions may occur between CO and intermediates of photochemical smog reactions. One intermediate is hydroxyl (OH) radical which can be produced by the photolysis of aldehyde. CO also reacts with other free radicals. CO undermines the self cleansing ability of the atmosphere by lowering the concentration of the hydroxy radical which is an important "detergent" because it reacts with nearly every gas molecule in the atmosphere. Without hydroxyl, the concentration of most trace gases would be much higher than those of today and the atmosphere as a whole would have totally different chemical, physical and climatic properties (Graedel and Crutzen 1989).

In the upper atmosphere, short wave ultraviolet radiation dissociates $CO_2$.

$$CO_2 + hv \rightarrow CO + O$$

$CO_2$ may be reformed by three body collision:

$$CO + O + M \rightarrow CO_2 + M$$

where M represents the third body.

CO is a relatively long lived substance. Various estimates of its residence time in the atmosphere vary from 0.1 to 5 years with an average of 0.2 years. The background levels of CO do not appear to be rising.

**Removal of CO from the Atmosphere**

This subject has been reviewed exhaustively by Jaffe (1970). The possible removal processes for CO are as follows:

1. CO in the lower atmosphere may possible migrate to the upper atmosphere by diffusion where it is oxidized to $CO_2$ in the presence of $NO_2$ or $O_3$ when it is subjected to high intensity ultraviolet radiation.

The oxidation of CO by the hydroxy radical (OH) in the troposphere has been suggested to be the major CO sink by some authors but calculations do not support this.

2.a. **The terrestrial sink.** This consists of plants and micro-organisms that can metabolize CO. CO in contact with soil may be oxidized to $CO_2$ or converted to methane ($CH_4$) by common methane producing soil organisms in presence of moisture according to the reaction:

$$4\,CO_2 + 2\,H_2O \rightarrow CH_4 + CO_2$$

In the presence of $H_2$, these bacteria can convert CO directly into methane and water:

$$CO + 3H_2 \rightarrow CH_4 + H_2O$$

b. The process of plant respiration may serve as a potential CO removal process. Plants may be able to metabolize CO to $CO_2$ which is then absorbed photosynthesis in the usual way. However, high concentrations of CO can be toxic to plant life.

3. **Biochemical sink.** CO binds to porphyrin type compounds that are widely distributed in plant and animal life. Permanent removal from atmosphere may depend on whether CO subsequently enters into some chemical process to form $CO_2$ when the porphyrin compound is degraded.

4. **Oceanic sink.** There is no evidence that oceans serve as a sink for CO. The solubility of CO in sea water is very limited. Swinnerton et al (1970) measured the CO content of the atmosphere and the surface water simultaneously at several points between Washington D.C. and Puerto Rico. The CO concentration of the surface water has 7 to 90 times (average 28 times) the theoretically calculated concentration of CO in the water if atmospheric CO was the only source. This points to the marine life in the oceans as a source of CO with transport of CO from the water to the atmosphere. The oceans release CO into the air which represents 5% of the estimated man-made CO, thus making the ocean the largest natural source of CO known.

5. **Rain water.** It is believed that the fresh water may remove some CO from the atmosphere. Rain water contains appreciable quantity of CO and run-off into the lakes and the rivers may account for additional removal.

## An Overview of the CO Cycle

Seiler (1974) has reviewed the cycle of atmospheric CO. My concept of the CO cycle is shown in Figure 3-3.

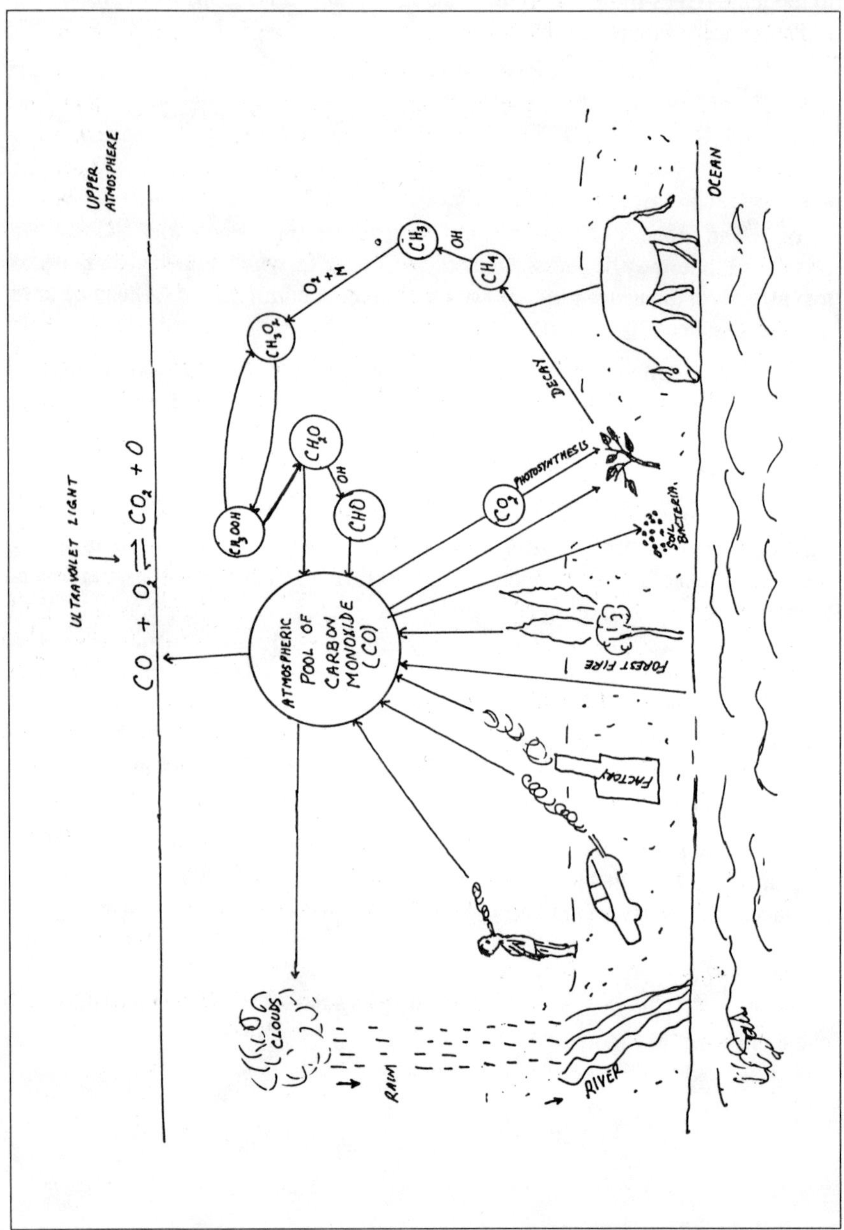


Fig. 3-3: The carbon monoxide cycle in the environments.

## Methods of Measuring CO in the Ambient Air

Three methods are routinely used for the estimation of CO in ambient air. These are:

1. **Continuous analysis method based on non-dispersion infrared spectroscopy** (NDIR). Any CO introduced into the sample cell will absorb radiation at the characteristic band centered at 4.67 um, causing the detector to produce an output signal proportional to the concentration of CO. The detection limit is about lm/m³ (0.87 ppm).
2. **Semicontinuous analysis method using gas chromatographic techniques.** In this CO is catalytically reduced to methane and passed through a flame ionization detector, the output signal of which is proportional to the concentration of CO in the air sample. It is particularly suitable when low concentrations of CO have to be measured with a high degree of specificity.
3. **Semiquantitative method employing detector tubes.** It is a simple method and can be used for estimating concentrations of CO above 5 mg/m³.

Other methods include catalytic oxidation, electrochemical analysis, mercury displacement, and the dual isotope technique. The details are described in a World Health Organization manual (WHO 1976).

An important part of any CO measurement procedure is the calibration technique. Many publications deal with this topic and the Deutsche Industrienormen Ausschuss (DIN) and the International Standard Organization (ISO) have special groups for establishing suitable calibration standards.

## Environmental Safety Standards

TABLE 3-4: ENVIRONMENTAL SAFETY STANDARDS FOR CARBON MONOXIDE

| | |
|---|---|
| *Public* | |
| EPA (Environmental Protection Agency | 1 mg/m³ (9 ppm) —maximum 8 h. |
| USA | 40 mg/m³ —maximum 1 h.<br>Neither to exceed more than once/yr |
| *Industrial* | |
| National Institute of Occupational Safety and Health, USA. | 35 ppm— 8 h. Time-weighted average. Ceiling 220 ppm. |
| American Conference of Government and Industrial Hygienists | Threshold limit value—50 ppm. Short-term exposure limit—15 min, time weighted average exposures that should not be exceeded. |
| Switzerland and Germany | 8 h. exposure to 30 ppm or 4% COHb* |
| Sweden | 35 ppm for 8 hr exposure. |

* According to Peterson and Stewart's formula (see Chapter 2) exposure to 18-33 ppm of CO for 8 hr will lead to COHb of 3-5% and exposure to 5-8 ppm for 24 hr to COHb of 2-3%.

Environmental safety standards for some countries are shown in Table 3-4. Mere setting of safety standards is not enough. Risk of exposure should take into consideration the dose-response curve and special risk groups such as the elderly and the children (Neuberger 1983).

In a survey of COHb levels of non-smokers, Stewart *et al.* (1974) found that 45% of them had COHb levels of more than 1.5% indicating that exposure to CO in excess of that permitted by Air Quality Standards was widespread.

## INDOOR CO CONCENTRATIONS

CO is widely generated indoors by heating, cooking and smoking. CO accumulation may also occur in other closed spaces as well. Indoor levels are generally unaffected by the outdoor CO levels. The extent of pollution in an enclosed space, from any source, can be calculated by the following formula (Tsunerari *et al.* 1985):

$$C = \frac{16^6 K(1\text{-}e^{Rt})}{RV}$$

Where

C = CO concentration in ppm after a given time (t) in hours.
R = Air changes/hour
V = volume of enclosed space (car or room in $m^3$)
K = Volume (in $m^3$) of CO released/hour (at atmospheric pressure)
e = base of the natural system of logarithms

As R or t increase, the factor ($1\text{-}e^{Rt}$) approaches 1 and equilibrium is reached. Where volume V is small, for instance in a car, CO concentration builds up rapidly:

$$\frac{K}{RV} = \text{Equilibrium concentration}$$

For a medium size care with a two liter engine capacity and a 0.05% concentration of CO in the exhaust leaking into the passenger compartment, a concentration of 0.08% will build up in about 20 minutes.

The equilibrium CO concentration is likely to be achieved for an approximately 30% level of CO concentration in the exhaust gases. Once this equilibrium is reached, it is maintained and no significant changes will occur subsequently.

If the atmospheric CO concentration is known, the rate of uptake by the blood of an adult at rest can be calculated by the formula:

$$\% \text{ COHb} = 3 \text{ X } \% \text{ CO in air X minutes of exposure}$$

Automobile related CO pollution may occur in closed garages and in passenger compartment of the car. Modern parking garages have good cross ventilation but in older facilities, the only air intake is from the polluted atmosphere of the streets.

Particularly high CO concentration areas are the underground service stations and toll booths. Parking attendants have high COHb levels and reported figures range from 5-22%. In a recent study of attendants and users of underground garages in Mainz, Germany, only slight increase in COHb was observed. All the safety regulations were followed in these garages. Half of the employees still experienced some complaints and the authors of this study (Schirmacher and Zander 1989) stated that deterioration of health was possible in high risk groups even at the slightly increased CIHb levels.

A further hazard area is within an airline terminal building due to idling of taxis and buses.

CO accumulation may occur in the passenger compartment of a car. The causes of this are:

1. Faulty exhaust system
2. Suction of car fumes from the car ahead in a traffic jam.
3. Open windows in a traffic tie-up inside a tunnel. The CO concentrations inside the car are then the same as in the air in the tunnel.

CO intoxication is most likely to occur in low income groups in the USA who require private transport but do not have the financial means to maintain the car. In one study (Williams 1985), 18.6% of the cars owned by Navajo Indians had compartmental CO levels exceeding 9 ppm.

Child passengers in a car are particularly vulnerable to high CO levels. The risk factors for fatal poisoning are: freezing temperatures, small body size, and sitting at rear (Zeller *et al.* 1984). Children lying at the rear sear of a station wagon may be affected while the front seat passengers may show no effects (Lacey 1981).

CO concentrations inside a house can rise due to several reasons. Overenthusiastic amateur draft-proofing is one cause. Other cheap methods of keeping warm such as burning paraffin or bottled gas in a small room can increase CO concentrations. Faulty gas heaters without a flue can lead to CO build up inside a house (Michaelson and Taudorf 1983).

The most frequent site of CO poisoning inside a home is the kitchen. Twenty years ago when coal gas contained, more than 30% CO, the number of deaths in the kitchen was much higher. Now-a-days natural gas contains 4-14% CO. It burns more efficiently and cleanly as compared to other forms of fuel but if the combustion is not complete it can be a source of CO pollution. Incomplete combustion of other fuels such as charcoal and wood can release CO which can be trapped inside a building if the chimney is clogged. Sofoluwe (1986) reported extremely high concentrations of CO in Nigerian dwellings when firewood was used for cooking. During the preparation of meals, average CO concentrations were reported to be over 1000 $mg/m^3$ (870 ppm) with peak levels as high as 3400 $mg/m^3$ (2960 ppm). Sidorenko *et al. (1970)* found an indoor level of CO of 32.3 $mg/m^3$ (28ppm) some two and a half hours after domestic combustion began, when there was no ventilation. This concentration dropped to 13.4 $mg/m^3$ (11.6 ppm) when ventilation was provided. The high incidence of CO poisoning in Korea is due to the widespread use of a unique Korean heating system called "Ondol" (56.4% of households). Ondol is similar to the European heating system of the Roman period called "Hypokaust" or "Kanal Heizung." Heated air by burning the coal briquettes passes through horizontal flues below the mud-plastered stone floor of the room and exhausts through the chimney located on the opposite side of the fireplace. The poor construction of the Ondol frequently leads to leakage of CO into the house (Kim and Yun 1969).

The living room in the house may also have a high CO level. Cox and Whichelow (1985) found that expired air breath CO in 21% of non-smokers, measured in their living rooms with heaters on, ranged form 6-42 ppm (corresponding to 1-7% COHb in blood). The cause of this was considered to be CO generated by the heating systems.

Excessive levels of CO have been found in ice-skating rinks where ice resurfacing machines are used. Levels as high as 348 mg/m$^3$ were found in such an arena after complaints of illness among children skating there were reported to the local health department (Johnson *et al.* 1975). Improperly regulated space heaters and faulty ice resurfacing machines can be the causes of high CO concentrations in these places.

Exhausts of many industrial plants, mills, and workshops contain CO. Risks are particularly high in blast furnace and coal mines. Explosives can emit as much as 60% CO.

## EPIDEMIOLOGY OF CO POISONING

In the United States, more than 3,800 persons die annually from CO poisoning and over 10,000 miss at least one day of work/yr because of sublethal exposure to CO (MMWR 31:529-531, 1982). In England and Wales, 1365 deaths from CO poisoning were reported in 1985 whereas only 475 admissions and 10 deaths were listed as due to CO poisoning in hospital records (Meredith and Vale 1988). In Korea, the incidence of CO poisoning in households with Ondol 5.4% to 8.4% as shown in a survey of 4 major cities (Cho *et al.* 1986).

Over the period 1965-1974, one million persons were reported as suffering from noxious coal briquette exhaust containing CO. During this period, 4304 persons were intoxicated with CO, and 300 of these died (Cole 1975). There are seasonal variations in CO poisoning in urban Korea. There are nine times as many cases of CO poisoning in December as in August. The incidence of CO poisoning correlates with ambient temperature and domestic heating but not with humidity (Kim 1985). In England, CO poisoning involving domestic heaters is more common in winter.

## CAUSES OF CO POISONING

Carbon monoxide is present universally but clinically manifest poisoning occurs only when critical levels are exceeded. Various causes for this are listed

TABLE 3-5
CAUSES OF CO POISONING

1. Automobile exhaust
2. Defective domestic appliances for heating and cooking
3. Suicidal inhalation of cooking gas or automobile exhaust.
4. Industrial plant exhausts
5. Mining accidents
6. Fires and smoke inhalation
7. Poisoning by methylene chloride (paint stripping solvent) due to its conversion to CO in vivo.
8. Cigarette smoking

in Table 3-5. Endogenous CO is unimportant because the values seldom exceed 3% COHb. The most important sources of CO poisoning are exogenous.

Most fatal cases of CO poisoning occur at home. Of the 340 deaths reported by the United States Consumer Safety Commission in 1982, 290 (80%) were due to defective gas appliances. Two patients have been reported to suffer myocardial infarction and pulmonary embolism after exposure to CO from defective natural gas water heater and furnace (Grace and Platt 1981). Twelve patients were reported to suffer from occult CO poisoning after exposure to kerosene space heaters (Fisher and Rubin 1982).

A kitchen is also the classical site of suicide by housewives, who close the kitchen door and lie on the floor with the gas turned on but not ignited. Another method of suicide is exposing oneself to the car exhaust with a running motor in a closed garage. A lethal concentration of CO can be reached in about 10 minutes. CO from the automobile exhaust is also a major cause of unintentional fatal poisoning in the United States. In 1980, it accounted for 611 (14%) of the nation's 4,331 unintended poisoning fatalities, 2.7 deaths per million population. In West Virginia with a predominantly rural population in low income category this incidence was nearly three times the national average. Baron *et al.* (1989) investigated the causes and found that most of the deaths occurred in older model vehicles with defective exhaust systems and while they were stationary.

Most of the deaths from car exhausts occur indoors but if the victim is in close proximity to the car, death can occur in the open air as well.

Dimaio *et al.* (1987) reported three suicidal deaths where the individuals were found lying in the open air with their faces in close proximity to the exhaust with the engine running.

Smoke contains CO and smoke inhalation injury is usually associated with CO intoxication. Wood and paper fires contain 12% CO. Firemen are particularly at risk from CO poisoning. COHb levels of 50% have been found in those dying within the first 12 hours of post-burn period, implicating CO as the main culprit (Zikria 1972).

It should be noted that the causes of CO poisoning listed in Table 3-5 correspond to the sources of environmental CO listed in Table 3-1 with the exception of methylene chloride which leads to CO poisoning because of hepatic metabolism (Stewart and Hake 1976).

## FACTORS WHICH AGGRAVATE CO POISONING

Factors which aggravate CO poisoning are listed in Table 3-6.

TABLE 3-6
FACTORS WHICH AGGRAVATE CARBON MONOXIDE POISONING

1. High CO concentrations in the atmosphere
2. Prolonged duration of exposure
3. Exercise in an atmosphere polluted with CO
4. Cigarette smoking in an atmosphere polluted by CO
5. Acute exposure to CO of persons with chronically high COHb levels
6. Exposure to CO in presence of hypoxia e.g. altitudes above 1500 m.
7. Simultaneous exposure to other toxic agents
8. Pregnancy
9. Old age
10. Pre-existing medical disorders:
    Myocardial ischemia
    Cerebral ischemia
    Anemia
    Thyrotoxicosis
    Fever
    Diabetes

**Exercise in an atmosphere with an increased CO content.** During exercise the gas-exchange is faster, cardiac output higher and more CO reaches the plasma in a shorter time. Exercise increases the oxygen requirement and causes relative hypoxia. Carbon monoxide absorption is accelerated 2-fold if the subject is walking and 3-fold if the subject is exercising strenuously.

**Level of CO in the air.** Exposure to 0.01% CO for one hour does not produce any symptoms but exposure to 0.1% may be fatal.

**Duration of exposure.** Long term exposure to low levels of CO may lead to slow saturation of the body with CO until the cytochromes are overwhelmed and the patient may go into a deep coma with low levels of circulating COHb. Short term exposure may cause a simple rise of COHb without significant evidence of CO toxicity. Although the severity of CO intoxication is greater with prolonged exposure, it does not involve accumulation of CO in the tissues (Sokal *et al.* 1984).

**Effect of high altitude on CO toxicity.** Several studies have shown that the toxicity of low concentrations of CO is aggravated at high altitudes. A concentration of CO of 10% in the air at 10,000 ft. reduces the oxygen carrying capacity of the Hb by 10.5%, thereby bringing the blood into a state of anoxemia (Heim 1939). Increment of 5-10%, in the level of COHb has been shown to produce a significant deterioration in the performance in flicker-fusion frequency test (FFT) at altitudes of 5000-6000 ft which normally do not affect FFt (Lilienthal *et al.* 1946). This test usually shows impairment, with mild hypoxia at 9000-12000 ft. McFarland *et al.* (1944) reported that CO, in smaller amounts than previously reported, may have a harmful effect, particularly at high altitudes. They used a visual discriminometer to show that smoking one cigarette, which increased COHb to 2%, caused impairment of the visual sensitivity.

**Effect of age on CO toxicity.** Newborn animals are more resistant than adults to the effects of hypoxic hypoxia and similarly to CO poisoning. Mice ranging in age from 1-150 days were exposed to 2000 ppm CO or 7.5% oxygen (hyposia). Two-

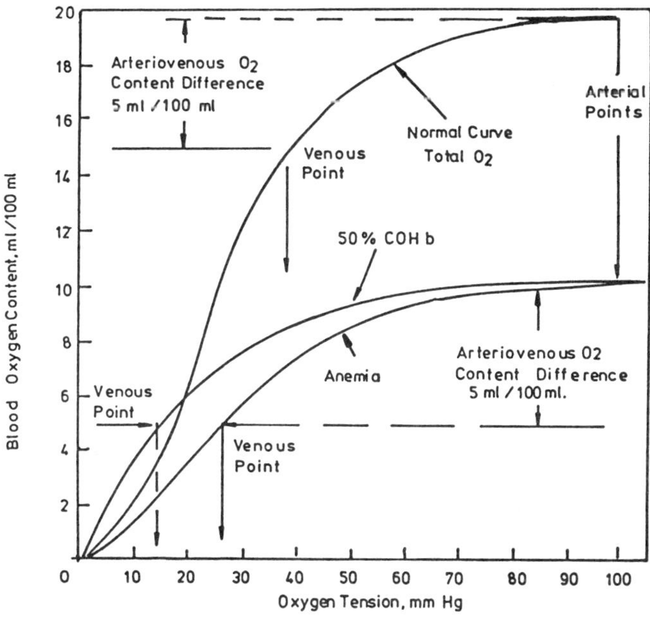


Fig. 3-4: Influence of anemia and CO poisoning on the relationship between oxygen tension and content. Arteriovenous oxygen difference of 5 ml/100 ml is assumed throughout. Normal curve is for hemoglobin of 14.4 gm/100 ml, resulting in venous $pO_2$ of 27 mm Hg, a low but not dangerous level. Curve of 50% COHb is based on total hemoglobin level (including COHb) of 14.4 gm/100 ml and gives a venous $pO_2$ of only 14 mm Hg, a level, that is dangerously low. Arterial $pO_2$ is assumed to be 100 mm Hg in all cases. *(Redrawn after Winter and Miller 1976).*

day mice were found to be more resistant to the lethal effects of both CO and hypoxia (Winson and Roberts 1978). The role of body temperature was also studied. Neonatal mice remained resistant to the lethal effects of CO until the ambient temperature reached 35°C, when the death rate of neonatal mice equaled that of adult mice in a 30°C environment.

**Anemia.** Anemia predisposes to CO poisoning as shown in Figure 3-4 and the accompanying legend.

**Carbon monoxide exposure in pregnancy.** Endogenous production of CO is doubled in women during the progesterone phase of the menstrual cycle as compared to the level during the estrogen phase or the level in men. During pregnancy, the rate of endogenous production increases even further but it declines rapidly during the post-partum phase. The cause of increased endogenous production of CO during pregnancy is not clear but there are known to be several contributing factors (Longo 1970). Up to 40% of the increase is accounted for by the increase in erythrocyte mass of the pregnant women. Progesterone may also stimulate the hepatic microsomal enzymes, thus contributing to the endogenous production of CO by the fetus assuming that it is equal to that of the neonate (Falstrom 1969).

Both CO and oxygen usually diffuse passively across the placenta in response to partial pressure gradients. Placenta may play an active role in the diffusion of CO (Burns and Gurter 1974). If we assume that the partial pressure of CO in the maternal blood is equal to the partial pressure of CO in the fetal blood, the steady state relation of fetal to maternal blood will depend on the relative affinity of oxygen for the fetal and the maternal blood as well as the relative affinities of CO as compared to that of oxygen. The normal steady state ratio of the fetal to the maternal COHb is 1:1, therefore, the fetal COHb concentration is about 10% to 15% higher than that of the mother.

**Simultaneous exposure to other toxins.** When an individual is exposed to two toxins simultaneously, the following types of effects may occur:

1. Independent. The effects of the two toxins may be totally unrelated and have different mechanisms.
2. Additive. The effects of the two toxins are similar and the combines effect is equal to the sum of the toxicities of the two agents.
3. Synergistic. One toxin acting by a different mechanism enhances the toxicity of the other, so that the total toxicity exceeds the sum of the individual toxicities of the two toxins.
4. One toxin may antagonize the effect of the other.

Interaction of CO with other toxins has been reviewed in detail by Pankow and Ponsold (1974). Important effects of simultaneous exposure of CO and other toxins are shown in Table 3-7.

TABLE 3-7
EFFECT OF OTHER TOXINS ON THE CO TOXICITY

| | Effect when given along with CO exposure | | | | |
|---|---|---|---|---|---|
| *Substance* | *Independent* | *Additive* | *Synergistic* | *Antagonistic* | *Remarks* |
| Carbon dioxide | | | yes | | see Chapter 2 |
| Nitrous oxide | | | yes | | Blocks Hb function (like CO) by met-hemoglobin formation. |
| Hydrocyanide | | | yes | | Quickened respiration leads to faster uptake of CO. |
| Sodium nitrite | | yes | | | |
| Sulfur dioxide | | yes | | | |
| Hydrogen peroxide | yes | | | | |
| Alcohol | yes | | | | |
| Trichlorethylene | | | yes | | Direct toxic effect on CNS cells |
| Benzol | | | | | |
| Carbon tetrachloride | | | | yes | Possible inhibition of liver oxidases (by CO) required for hepatotox-icity |
| Iodoacetate | | | | yes | |
| Barbiturates | | | | yes | Cellular protective effect of barbiturates by reduction of metabolism |
| Iodine acetate | | | | yes | |

## OCCUPATIONAL HAZARD OF CO EXPOSURE

Many occupational groups are subject to high CO exposure. These are listed in Table 3-8.

Traffic policemen have also raised COHb levels but the risk varies according to the number of hours on duty (usually not more than a few hours at a time) and the length of off-duty breaks to recuperate. Most of the reports in literature describe only a slight increase in COHb in non-smoking policemen. The values of COHb in smoking policemen are much higher.

Direct measurement of COHb levels in firefighters engaged in prolonged firefighting indicated that 10% had values exceeding 10% (Gordon and Rogers 1969). CO levels were measured in city fires in Los Angeles and values as high

TABLE 3-8

POTENTIAL OCCUPATIONAL EXPOSURES TO CARBON MONOXIDE

*(Reproduced from Shepherd, 1983, by permission)*

(a) Exposure to coal and coke fires
Bakeries—boiler rooms—brick burning—cooks—firemen—furnace starters—glass blowers—heat treaters—locomotive drivers and stokers.

(b) Metal foundries and refining
Blast furnace—brass foundry—lime manufacture, reduction of metallic oxides—nickel refining and smelting—steel manufacture—welding.

(c) Exposure to gasoline exhaust
Vehicle operators—garage workers—traffic and tunnel police, vehicle inspectors—marine surveyors—longshoremen and warehousemen, forklift operators—street workers—aircraft pilots

(d) Exposure to manufactured gas
Gas vehicle operators—production of coal, water, and producer gas—petroleum industry—sewer repairs—coal tar distilling

(e) Exposure to chemical processes
Manufacture of acetic acid—acetylene—alcohols—ammonia—beer—methanol, organic chemicals—oxalic acid—paper manufacture—paint solvents—phthallic anhydride—soda—zinc oxide

(f) Miscellaneous
Graphic art—tending of livestock—space exploration—submarine operation—cigarette smoke

as 1000 ppm were observed (Barnard and Weber 1979). Repeated exposure to high concentrations of CO may explain the unusually high incidence of cardiovascular disease among firemen. Operators of gasoline powered machines are exposed to CO even outdoors. Huff and Kardon (1989) reported the case of mild CO poisoning in a teenager splitting wood outdoors with a gasoline powered hydraulic machine.

Many of the studies of COHb in workers cannot be evaluated properly due to the variations in the smoking habits of the workers. Smokers as a whole, tend to have higher COHb levels than non-smokers.

## REFERENCES

Barnard J, Weber JS (1979) Carbon Monoxide: A hazard to firefighters. Arch Environmental health 34:255-257.

Baron RC, Backer RC, Sopher IM (1989) Unintentional deaths from carbon monoxide in motor vehicle exhaust: West Virginia. Am J Pub health 79:328-330.

Burns B, Gurtner GH (1973) A specific carrier for $O_2$ and CO in the lung and placenta. Drug Met Disp 1:374.

Cho SH, Lee DH, Yun DR (1986) Incidence of carbon Monoxide Intoxication J Korean Med Assoc 29:1233-1240.

Cole PV (1975) Comparative effects of atmospheric pollution and cigarette smoking on carboxyhemoglobulin levels in men. Nature 255: 699-701.

Cox BD, Whichelow MJ (1985) Carbon monoxide levels in the breath of smokers and non-smokers: effects of domestic heating systems J Int Epidemiol Commun Health 39:75-78.

Di Maio VJM, Dana SE (1987): Deaths caused by carbon monoxide poisoning in an open environment (outdoors) J Forensic SC 32: 1794-95.

Falstrom FP (1969) On the endogenous formation of carbon monoxide in full-term newborn infants. Acta Pediatr Scand Suppl 189:1.

Fisher J, Rubin KP (1982) Occult carbon monoxide poisoning. Arch Int Med 142:1270-1271.

Grace TW, Platt FW (1981) Subacute carbon monoxide poisoning: another great imitator. JAMA 246: 1698-1700.

Graedel TE, Crutzen PJ (1989) The changing atmosphere. SC Am 261: 28-36.

Greenpeace (1988) Stau im Gotthardt-Tunnel. LNN April 5, 1988.

Goldsmith JR (1970) Contribution of motor vehicle exhaust, industry, and cigarette smoking to community carbon monoxide exposures. Ann NY Acad SC 174:122-134.

Goldman A, Mucray DG, Mucray FH et al (1973) Vertical distribution of CO in the atmosphere. J Geophysical Res 78:5273-5283.

Gordon GS, Rogers RL (1969) A report of medical findings of project carbon monoxide. Washington, DC, Merkle Press.

Huff JS, Kardon E. (1989) Carbon monoxide toxicity in a man working outdoors with a gasoline powered hydraulic machine.
NEJM 320:1564.

Hwang DH (1969) An epidemiological study on carbon monoxide poisoning in Seoul, Korea. In: Kim ID, Yun DR: Carbon Monoxide Poisoning, Seoul, New Medical Publication, pp 108-113.

Heim JW (1939) The toxicity of carbon monoxide at high altitudes.
J Aviation Med 10:211-215.

Isler R, Keller A, Wenner H-U (1989) Luftschadstoffbelastung von Fussgängern, Joggern, Velos und Autofahren in der Stadt Zürich.
UMWELTTECHNIK 23:61-63.

Jaffe LS (1970) Sources, characteristics, and fate of atmospheric carbon monoxide.
Ann NY Acad Sc 174:76-88.

Johanson CJ, Moran JC, Paine SC et al (1975) Abatement of toxic levles of carbon monoxide in Seattle ice rinks.
Am J Pub health 65:1087-1090.

Kim ID, Yun DR (1969) Carbon Monoxide poisoning, Seoul, New Medical Publication, p. 290.

Kim YS (1985) Seasonal variation in carbon monoxide poisoning in urban Korea.
J of Epidemiology and Community Health 39: 79-81.

Lacey DJ (1981) Neurologic sequelae of acute carbon monoxide intoxiation.
Am J Dis Child 135:145-147.

Lilienthal JL, Fugitt CH (1946) The Effect of Low Concentrations of Carboxyhemoglobin on the "Altitude Tolerance" of Man.
Am J Physiol 145: 359-364.

Longo LD (1970) Carbon monoxide in the pregnant mother and fetus and its exchange across the placenta.
Ann NY Acad Sci 174:313.

McFarland RA, Roughton FJW, Haperin MH et al (1944) The effects of carbon monoxide and altitude on visual thresholds.
J Aviat Med 15 (1944) 381-394.

Michaelson KF, Taudorf K (1983) Danger of gas water heaters.
Lancet 1:229.

Meredith T, Vale A (1988) Carbon Monoxide Poisoning.
BMJ 296:77-79.

Neuberger M. (1983) New Approaches to Risk Estimation of Air Pollutants.
Vienna, Facultas Verlag, p. 8.

Newell RE, Reichle HG, Seiler W (1989) Carbon monoxide and the burning earth. Sc Am 261:58-64.

Pankow D, Ponsold W (1974) Kombinationswirkung von Kohlenmonoxid mit anderen biologisch aktiven Schadfaktoren auf den Organism. Z Gesamt Hyg und ihre Grenzgebiete (DDR) 9:561-571.

Robinson E, Robbins RC (1970) Atmospheric background concentrations of carbon monoxide Ann NY Acad Med 174:89-95.

Schirmer P, Zander R (1989) CO-Belastung bei Personal und Benutzern von Tiefgaragen. Arbeitsmed Sozialmed Praventivmed 24:5-8.]

Seiler W (1974) The cycle of atmospheric CO. Tellus 26:116-135.
Stewart RD, Hake CL (1976) Paint remover hazard. JAMA 235:398-401.

Sidorenko GI, Lampert FF, Pinigina MA et al (1970) New data on the concentration of the products of incomplete combustion of domestic gas in the air of premises provided with a gas supply. In: Problems of current interest in hygiene, Moscow, Akademija Medicinskij Nauk USSR, Institute of Genetics and Community Hygiene. pp 137-142 (in Russian).

Sofoluwe GO (1968) Smoke pollution in dwellings of infants with bronchopneumonia. Arch Environ Health 16:670-672.

Sokal JA, Majka J, Palus J (1984) The content of carbon monoxide in the tissues of rates intoxicated with carbon monoxide in various conditions of acute exposure. Arch Toxicol 56: 106-108.

Stewart RD, Baretta ED, Platte LR et al (1974) Carboxyhemoglobin levels of American blood donors. JAMA 229:1187-1195.

Swinnerton JW, Linnenbom VJ, Lamontagne RA (1979) Distritution of carbon monoxide between the atmosphere and the ocean. Ann NY Acad 174:96-101.

Tsunenari S, Kanda M, Yonemitus K et al (1985) Suicidal carbon monoxide inhalation of exhaust fumes. The American J of Forensic Med and Path 6:233-239.

Wanner HU (1984) Auswirkungen der Luftvermutzung auf die Gesundheit. Schweiz. Ärteschr 65:1523-28.

Wanner HU (1989) Personal communication.

Wharton M, Bistowish JM, Hutcheson RH et al (1989) Fatal carbon Monoxide poisoning at a motel. JAMA 261:1177-1178.

WHO (1976) Selected Methods of Measuring Air Pollutants, Geneva, WHO.

WHO (1979) Environmental health Criteria 13: Carbon Monoxide. Geneva, World Health Organization.

Williams RL (1985) Vehicular carbon monoxide screening: identification in a cross-cultural setting of a substantial public health risk factor. Am J Pub Health 75:85-86.

Winston JM, Roberts RJ (1978) Influence of Increasing Age on Lethality Induced by Carbon Monoxide or Hypoxic Hypoxia. Biol. Neonate 34: 199-202.

Zeller WP, Miele A, Suarez C et al (1984) Accidental carbon monoxide poisoning. Clin Pediatr 23:694-695.

Zikria BA, Wistin GC, Chodoff M et al (1972) Smoke and carbon monoxide poisoning in fire victims. J. Trauma 12:641-7.

# Chapter 4

## SMOKING AND CO POISONING

### INTRODUCTION

Several diseases are recognized to be more common among smokers. These include cancer of the lungs, breast and larynx; chronic obstructive pulmonary disease, stroke, heart disease and peptic ulcer. Mortality of smokers is 70% higher than that of a comparable non-smoking population (Hammond *et al.* 1958, Mulcahy *et al.* 1970, Strong and Richards 1976).

Smoking is the commonest and most underestimated cause of chronic non-fatal CO intoxication. More than 3,800 compounds have been identified in tobacco smoke. The major constituents which are measured and implicated in the pathogenesis of various diseases are: tar, nicotine and CO. Tar constituents are considered to be carcinogenic. In studying the effects of smoking in human subjects, it is difficult to sort out the effects of nicotine from those of CO. Studies with ingestion of pure nicotine show that it has a euphoric effect and improves mental activity. Acute nicotine infusions in the awake rat have been shown to increase the local cerebral glucose utilization in several areas of the brain (Grunwald et al. 1988). CO, on the other hand, is a central nervous system depressant. From this one may anticipate that the acute effects of smoking in a non-smoker may facilitate mental activity whereas chronic smoking with CO accumulation may depress mental activity.

Endurance time during exercise is decreased by 20% while smoking cigarettes whereas it is decreased by only 10% if an equivalent amount of CO is breathed. Klausen *et al.* (1983) showed that decrease in $VO_{2\,max}$ during smoking is due to CO saturation of the blood (decrease in the oxygen capacity of the blood), whereas a decrease in the endurance time during smoking is a combined effect of CO saturation and an increased cost of breathing caused by smoke particles.

Nicotine has also been implicated in the deleterious effect of smoking on the vascular system. Smoking is recognized as a risk factor in cardiovascular and cerebrovascular diseases. Both nicotine and CO play a role in platelet aggregation and enhanced atherosclerosis. Nicotine is responsible for the increased blood pressure and heart rate after smoking. CO has a negative ionotropic effect and decreases stroke volume. Usually there is no significant change in cardiac index after smoking. However, the cardiac index is significantly decreased after inhaling 150 ppm of CO because the inhaled CO significantly decreases the stroke volume but does not affect the heart rate. Therefore, patients with myocardial insufficiency develop angina sooner after exercise following cigarette smoking for at least two reasons (Aronow *et al.* 1974):

1. Increased myocardial oxygen demand caused by nicotine.
2. Reduced oxygen delivery to the myocardium whether or not nicotine is present.

## SMOKING AND COHB LEVELS

There are numerous reports in the literature showing that smokers have raised COHb levels. Using a COHb level of 2% as a cut off point, 81% of the smokers have been shown to have raised COHb levels as compared to 1% of non-smokers in similar environmental conditions (Wald *et al.* 1981). Smokers have higher COHb levels than non-smokers when exposed to similar levels of CO in the environment (Table 4-1). The concentration of COHb depends on several factors such as the type of tobacco, length of the cigarette, type of filter, speed of smoking, number of cigarettes smoked and the pattern of smoke inhalation. Inhalation of the smoke of one cigarette raises COHb level by 1%.

TABLE 4-1
COMPARISON OF COHB LEVELS OF SMOKERS WITH THOSE OF NON SMOKERS UNDER DIFFERENT CONDITIONS OF EXPOSURE

*(Reproduced from Lawther and Commins 1970)*

| | COHb% | | | | | | | |
|---|---|---|---|---|---|---|---|---|
| | Smokers | | | | Nonsmokers | | | |
| | No. | Range | Mean | S.D. | No. | Range | Mean | S.D. |
| Controls ("unexposed") | | | | | | | | |
| Exposure to traffic in streets | 165 | (0.4- 9.5) | 3.5 | 2.1 | 155 | (0.0-2.5) | 0.8 | 0.4 |
| (i) Police on foot | 29 | (1.3- 6.6) | 3.6 | 1.3 | 11 | (0.6-3.8) | 1.9 | 0.4 |
| (ii) Others on foot | - | - | - | - | 5 | (0.6-3.8) | 1.2 | 0.4 |
| (iii) Others cycling | - | - | - | - | 6 | (1.5-2.1) | 1.7 | 0.3 |
| (iv) Others in cars | - | - | - | - | 8 | (0.8-3.3) | 1.8 | 0.8 |
| Exposure to traffic in confined spaces | | | | | | | | |
| (i) Garages | 22 | (2.3-8.6) | 5.0 | 1.8 | 21 | (0.7-5.9) | 2.4 | 1.3 |
| (ii) Customs sheds, car ferries, tollbooth operators | 34 | (0.8-8.6) | 4.2 | 2.2 | 23 | (0.8-2.8) | 1.3 | 0.5 |
| Exposure to petrol engine exhaust in greenhouses | 5 | (1.8-31.9) | 14.1 | 11.4 | 1 | - | 8.1 | |
| Firemen | 36 | (0.6-15.1) | 5.1 | 3.2 | 32 | (0.4-8.8) | 3.2 | 2.0 |

TABLE 4-2
COHB LEVELS ACCORDING TO THE NUMBER OF CIGARETTES CONSUMED

| No. of cigarettes per day | Median COHb concentrations (%) | | Authors |
|---|---|---|---|
| | Smokers | Non-smokers (control) | |
| 3 to 4 | 3.6 | 2.5 | Pankow et al 1976 |
| >5 | 4.1 | 1.6 | Effenberger et al 1957 |
| 5 to 10 | 5.1 | 2.5 | Pankow et al 1976 |
| 11 to 20 | 6.6 | 2.5 | Pankow et al 1976 |
| 11 to 20 | 6.1 | 1.3 | Castledon and Cole 1975 |
| >20 | 6.7 | 1.3 | " |
| >20 | 7.3 | 2.5 | Pankow et al 1976 |
| >20 | 5.0 | 0.9 | Butt et al 1974 |
| 10 to 40 | 5.9 | 1.3 | Goldsmith 1974 |
| >40 | 6.9 | 1.3 | " |
| 30 to 80 | 11.6 | 0.6 | Smith and Landau 1978 |

The average concentration of COHb in smokers is approximately 5%. Results of various studies on this subject are shown in Table 4-2.

Haynes *et al.* (1983) recorded a mean COHb concentration of 8.26 ± 1.08 in their subjects 15 minutes after smoking the last of 20 cigarettes without interruption. When tobacco-free cigarettes were smoked by these subjects, the COHb levels were even higher (13.08 ± 1.69).

Inhaled smoke of most cigarettes contains 400 ppm of CO. This is eight times higher than the industrial standard set up by the Occupational Safety and Health Administration for an eight hour daily exposure (50 ppm). Cigar smoke has a somewhat higher content of CO. Low tar and nicotine containing cigarettes release 60 ppm of CO from burning of the paper and can raise COHb levels to 12.55 (Kanzler *et al.* 1983). However, other investigators have shown that switching to low tar/low nicotine cigarettes lowers COHb levels in smokers of regular cigarettes. Another study (Benowitz *et al.* 1986) showed that lower-yield cigarettes do not affect the levels of CO in smokers unless the yield is ultra low (1 mg of tar).

Most smokers are addicted to nicotine; they tend to compensate for their low yield cigarettes by smoking them in such a way as to optimize the intake of nicotine and incidentally take in more CO (Benowitz 1989). Zacny *et al.* (1986) have observed that the smokers, by blocking the vents of the "ultra low yield" cigarettes, exposed themselves to CO concentrations double that of non-blocked cigarettes (8.96 ppm versus 4.32 ppm). CO exposure has been found to be substantially greater after puffing full length cigarette rods than after short cigarette rods (Nemeth-Coslett and Griffiths 1985). Nil *et al.* (1986) found that smoker with high CO absorption rates were more addicted to nicotine and exhibited more intensive patterns of puffing and respiratory inhalation. Due to various smoking styles and individual responses, there is no uniformity of opinion on this subject. Smokers with relatively high COHb levels show a greater increase after smoking a single cigarette than individuals with low COHb (Burling *et al.* 1985).

Marijuana cigarette smoking has been shown to raise venous COHb by 40% in non-smokers while the mean rise in expired CO was 100% (Hecht and Bogt 1985). Absorption of CO from smoking one marijuana "joint" is four times than that from smoking a cigarette.

**Passive smoking**

This term is used to describe the involuntary inhalation of smoke from active smokers in a closed environment. Environmental tobacco smoke is derived from two sources; mainstream smoke and sidestream smoke. Mainstream smoke is the complex aerosol mixture inhaled by the smokers. Sidestream smoke is the aerosol emitted directly into the surrounding air from the lit end of a smouldering tobacco product. The two types of smoke share similar components but sidestream smoke has a higher concentration of CO than the mainstream smoke (Fielding and Phenow 1988). The high CO content of the sidestream smoke is due to greater production of it in the oxygen-deficient cone during smouldering than during puff-drawing. After passing through the hot cone, most of the CO is oxidized to $CO_2$, probably because of the high temperature gradient and sudden exposure to air. CO concentrations vary linearly with the rate of cigarette burning in passive smoking (Hoegg 1972) and values as high as 13% have been reported (Kachulis 1981). Under experimental conditions, simulating heavy pollution with sidestream smoke (CO, 20 m/liter or 20 ppm), the particulate tobacco smoke inhaled by involuntary smokers were equivalent to one half ton one cigarette per day (Hugod *et al.* 1978). In one study, rise of COHb in passive smokers was 1% as compared to 0.7% rise of COHb in active smokers per cigarette smoked (Russell *et al.* 1973). The amount of nicotine taken in by passive smokers is, however, is one-third to one-tenth of that taken in by the active smokers (Jarvis *et al.* 1983).

The committee on passive smoking of the National Research Council of USA (1986) has published a review of the environmental tobacco smoke (ETS). Measurements of concentration of CO in ETS are shown in Table 4-3. CO and COHb levels in non-smoking individuals are shown in Table 4-4.

TABLE 4-3

MEASURED CONCENTRATION OF CO IN ENVIRONMENTAL TOBACCO SMOKE

*From: Environmental Tobacco Smoke, National Academy Press, Washington, 1986, by permission*

| | | | Carbon Monoxide Concentrations, ppm | | | | |
|---|---|---|---|---|---|---|---|
| | | | | | Nonsmoke Controls | | |
| Location | Tobacco Burned | Ventilation | Mean | Range | Mean | Range | References |
| Rooms | — | — | — | 4.3-9 | 2.2 ± 0.98 | 0.4-4.5 | Coburn et al., 1965 |
| Train | 1-18 smokers | Natural | — | 0-40 | — | — | Harmsen and Effenberger, 1957 |
| Submarines | 157 cigarettes/day | Yes | <40 | — | — | — | Cano et al., 1970 |
| (66 $m^3$) | 94-103 cigarettes/day | Yes | <40 | — | — | — | |
| 18 military aircraft | — | Yes | <2-5 | — | — | — | U.S. Department of Transportation, 1971 |
| 8 commercial aircraft | — | Yes | <2 | — | — | — | U.S. Department of Transportation, 1971 |
| Rooms | — | — | — | 5-25 | — | — | Porthein, 1971 |
| 14 public places | — | — | <10 | — | — | — | Perry, 1973 |
| Ferry boat | — | — | 18.4 ± 8.7 | — | 3.0 ± 2.4 | — | Godin et al., 1972 |
| Theater foyer | — | — | 3.4 ± 0.8 | — | 1.4 ± 0.8 | — | Godin et al., 1972 |
| Intercity bus | 23 cigarettes | 15 changes/h | 32 | — | — | — | Seiff, 1973 |
| | 3 cigarettes | 15 changes/h | 18 | — | — | — | |
| 2 conference rooms | — | 8 changes/h | — | 8 (peak) | 1-2 | — | Slavin and Hertz, 1975 |
| Office | — | 236 $m^3$/h | — | <2.5-4.6 | — | — | Harke, 1974 |
| | — | Natural | — | <2.9-9.0 | — | — | — |
| Automobile | 2 smokers | Natural | — | 42 (peak) | — | 13.5 (peak) | Harke and Peters, 1974 |
| | (4 cigarettes) | Mechanical | — | 32 (peak) | — | 15.0 (peak) | |
| 9 night clubs | — | Varied | 13.4 | 6.5-41.9 | — | — | Sebben et al., 1977 |
| 14 restaurants | — | — | 9.9 ± 5.5 | — | 9.2 (outdoor) | 3.0-35 | Sebben et al., 1977 |
| 45 restaurants | — | — | 8.2 ± 2.2 | 7.1 ± 1.7 | — | — | Sebben et al., 1977 |
| 33 stores | — | — | 10.0 ± 4.2 | 11.5 ± 6.5 | 11.5 ± 6.5 | — | Sebben et al., 1977 |
| 3 hospital lobbies | — | — | — | 4.8 | — | — | Sebben et al., 1977 |
| 6 coffee houses | Varied | — | 2-23 | — | — | — | Badre et al., 1978 |
| Room | 18 smokers | — | 50 | — | — | — | Badre et al., 1978 |
| Hospital lobby | 12-30 smokers | — | 5 | — | — | — | Badre et al., 1978 |
| 2 train compartments | 2-3 smokers | — | — | 4-5 | — | — | Badre et al., 1978 |
| Automobile | 3 smokers | Natural, open | 14 | — | — | — | Badre et al., 1978 |
| | 2 smokers | Natural, closed | 20 | — | — | — | |
| 10 offices | — | — | 2.5 ± 10 | 1.5-1.0 | 2.5 ± 1.0 | 1.5 ± 4.5 | Chappell and Parker, 1977 |
| 15 restaurants | — | — | 4.0 ± 2.5 | 1.0-9.5 | 2.5 ± 1.5 | 1.0-5.0 | Chappell and Parker, 1977 |
| 14 night clubs and taverns | — | — | 13.0 ± 7.0 | 3.0-29.0 | 3.0 ± 2.0 | 1.0-5.0 | Chappell and Parker, 1977 |
| Tavern | — | Artificial | 8.5 | — | — | — | Chappell and Parker, 1977 |
| | — | None | — | 35 (peak) | — | — | |
| Office | — | Natural, open | 1.0 | 10.0 (peak) | — | — | Chappell and Parker, 1977 |
| Restaurant | — | Mechanical | 5.1 | 2.1-9.9 | 4.8 (outdoors) | — | Fischer et al., 1978 |
| Restaurant | — | Natural | 2.6 | 1.4-3.4 | 1.5 (outdoors) | — | Fischer et al., 1978 |
| Bar | — | Natural, open | 4.8 | 2.4-9.6 | 1.7 (outdoors) | — | Weber et al., 1976 |
| Cafeteria | — | 11 changes/h | 1.2 | 0.7-1.7 | 0.4 (outdoors) | — | Weber et al., 1976 |
| 44 offices | — | — | 1.1 | 6.5 (max) | — | — | Weber, 1984 |
| 25 offices | — | — | 2.78 ± 1.42 | — | 2.59 ± 2.33 | — | Szadkowski et al., 1976 |
| Tavern | — | 6 changes/h | 11.5 | 10-12 | 2 (outdoors) | — | Cuddeback et al., 1976 |
| Tavern | — | 1-2 changes/h | 12.0 | 3-22 | — | — | Cuddeback et al., 1976 |

Time-weighted average (TWA) of carbon monoxide, 50 ppm (55 mg/$m^3$). TWA = average concentration to which worker may be exposed continuously for 8 h without damage to health (National Institute for Occupational Safety and Health, 1971).

TABLE 4-4
CARBON MONOXIDE AND CARBOXYHEMOGLOBIN LEVELS IN NONSMOKING INDIVIDUALS

*Experimental Studies (Controlled Chambers)*

| Study | No. of Cigarettes/h/10 m³ | No. of Subjects | CO, ppm[a] | Carboxyhemoglobin | |
|---|---|---|---|---|---|
| | | | | Control | Change |
| Anderson and Dalhamm, 1973 | 3.1 | — | 4.5 | 0.3 | 0 |
| Dahms et al., 1981 | — | 10 | 15–20 | 0.6 | +0.4 |
| Harke, 1970 | 3.9 | 7 | 30 | 0.9 | +1.2 |
| Huch et al., 1980 | 2.3 | 12 | — | 1.3 | +0.5 |
| Hugod et al., 1978 | 2.5 | 10 | 20 | 0.7 | +0.9 |
| Pimm et al., 1978 | 2.4 | 10 | 24 | 0.5 | +0.3 |
| | 2.4 | 10 | 24 | 0.7 | +0.2 |
| Polak, 1977 | 6.7 | 15 | 23 | 2.0 | +0.3 |
| Russell et al., 1973 | 15.1 | 12 | 38 | 1.6 | +1.0 |
| Seppänen and Uusitalo, 1977 | 3.8 | 28 | 16 | 1.6 | +0.4 |
| Srch, 1967 | 50 | — | 90 | 2 | +3 |

*Observational Studies*

| Study | Subjects/Exposure | No. of Subjects | Nonexposed: Exposed | |
|---|---|---|---|---|
| | | | Carboxy-hemoglobin, % | CO Ex-pired, ppm |
| Foliart et al., 1982 | Flight attendants/8 h | 6 | 1.0:0.7 | |
| Jarvis et al., 1983 | Normal/public house for 2 h | 7 | | 4.7:10.6 |
| Lightfoot, 1972 | Normal/submarine | | —:1.0 | |
| Wald et al., 1981 | Participants in health screening program | 6,641 | | |
| Jarvis et al., 1984 | Normal/self report | 10 | 0.9:0.8 | 5.7:5.5 |
| Seppänen and Uusitalo, 1977 | Restaurant for 5 h (CO:2.5–15 ppm) | 47 | 2.1:2.1 | |
| | Office for 8 h (CO:2.5 ppm) | 15 | 2.3:2.3 | |

[a]Carbon monoxide (CO) measured as a proxy to indicate the concentration of ETS in the chamber.

*From: Environmental Tobacco Smoke.*
*Published by National Academy Press, Washington, D.C. 1986.*

CO is introduced by smoking into the completely sealed environments of the nuclear submarines (up to 100 ppm). Improvements in removal equipment have brought these levels down to 8-10 ppm in the Polaris submarines. This should not produce a COHb level of more than 1.6% in the non-smoking individuals. A survey on a Polaris patrol of the British Royal Navy showed COHb levels of 1.1% in non-smokers, 8% in twenty-a-day smokers, and over 18% in heavy smokers (Lightfoot 1972). This study shows that low ambient CO concentrations can have an additive effect on the relatively high COHb concentrations produced by smoking. Similar cumulative effect of smoking and occupational exposure to CO has been shown in other studies (Butt *et al.* 1974, Levy *et al.* 1976). COHb values in smokers before work are about double the values in non-smokers after work (Seppänen and Uusitalo 1977).

In a recent study of measurement of exhaled CO in flight attendants, who worked in the smoking section, no significant increase was found after a flight lasting five hours (Duncan and Greaney 1984).

## SMOKING AND THE CARDIOVASCULAR SYSTEM

Smoking as a risk factor for cardiovascular and cerebrovascular diseases is documented in detail elsewhere (Jain 1990). Of the many constituents of smoke, nicotine and CO appear most likely to have a role in the pathogenic effects of smoking on the cardiovascular system (Wald 1975). Ischemic heart disease is 2 to 5 times more prevalent in the smokers than in the non-smokers.

### Atherosclerosis and blood cholesterol

Among various causes of atherosclerosis, smoking is considered to be the number one preventable cause (Fischer and Richards 1986). Both nicotine and CO may be the culprits. CO-induced hypoxia of the vessel wall is a factor in atherosclerosis (Toping 1977).

Astrup (1973) proposed that the causal relationship between smoking and atherosclerosis stems from the abnormal levels in the blood of smokers. Wald *et al.* (1973) found that the risk of developing atherosclerotic disease correlates with the COHb levels of the smokers. Individuals in the age group 30-69 years with COHb levels above 5% were found to be 21 times more likely to be affected by atherosclerosis than those with COHb levels below 3%. Effeney (1987) studied the effect of nicotine and CO components of smoking on $PGI_2$ (a platelet aggregation inhibitor) production in the rabbit heart. Nicotine reduced $PGI_2$ production and because it inhibits a protective mechanism, it may be responsible for the initiation of the atherosclerotic plaque. Carbon monoxide leads to relative hypoxia of the endothelial cells of the vessel wall, which may stimulate $PGI_2$ production. However, CO can also have a direct toxic effect on the endothelium, which may accelerate the atherosclerotic process. There is increased fibrinogen uptake by the arterial wall in smokers. The atherogenic effects of this may involve two mechanisms; CO increases the permeability of the vessel wall whereas nicotine impairs fibrinogen clearance (Allen *et al.* 1984).

Tobacco smoke has an atherogenic impact on plasma lipids and lipoproteins (Sarma *et al.* 1975). In the Berlin-Bremen Study (Dwyer *et al.* 1988), longitudinal changes (over two years) in serum cholesterol were studied among adolescents exposed to low level smoking (CO - 9 to 13.5 ppm). There was a significant decrease of HDL-C smokers as compared to non-smokers thus suggesting the atherogenic potential of smoking. Autopsy studies have shown that 50% more of the intimal surface of the coronary arteries and the abdominal aorta are involved with atheromatous plaques in heavy smokers are compared to non-smokers (Strong and Richards 1976).

**Effect on hemorheology**

Increased blood levels of COHb in smokers are associated with a decreased index of erythrocyte deformability (Gelmini *et al.* 1985). Hypoxia induced by chronic CO intoxication resulting from smoking increases the Hb content of the blood (Pankow *et al.* 1976). Vascular damage can be caused by increased blood viscosity induced by smoking.

The importance of blood platelets in the pathogenesis of vascular disease is well known. Several authors suggest that smoking increases platelet activity (Levine 1973, Schmidt and Rasmussen 1982, Davis and Davis 1979, Bierenbaum *et al.* 1978, Grignani *et al.* 1978). The exact mechanism is not well understood but it is unlikely to a direct effect of nicotine (Pfueller *et al.* 1988).

**Cardiomyopathy**

Experiments in rabbits have shown that both factors of cigarette smoke (CO and nicotine) participate in the metabolic injury of mitochondria (Gvozdjakova *et al.* 1984). Long term cigarette smoking causes considerable metabolic and morphological alterations of heart muscle which can be characterized as smoke cardiomyopathy. The association of smoking with cardiomyopathy is separate from its association with coronary heart disease (Hartz *et al.* 1984).

**Cerebrovascular disease**

Cigarette smoking has been found to be a risk factor for stroke (Wolf *et al.* 1988). A positive association between cigarette smoking and carotid atherosclerosis has been reported (Bogousslavsky *et al.* 1985, Crouse *et al.* 1987, Passero *et al.* 1987, Ford *et al.* 1985). Abstention from cigarette smoking has been shown to reduce the thickness of carotid plaques (Tell *et al.* 1989) and improves cerebral perfusion among elderly chronic smokers (Rogers et al. 1985). Long term smoking may reduce rCBF, not by atherosclerosis, but by means of hypocapnia resulting from latent small airway obstruction (Yamashita *et al.* 1988).

Patients with ischemic stroke due to extra- or intracranial-vascular disease are at greater risk from stroke than previously reported (relative risk of 5.7%) whereas those with stroke due to cardiac emboli have no excess risk associated with smoking (Donnan *et al.* 1989).

Atherosclerotic changes in the internal auditory artery have been suggested as an explanation of the poorer extra-high frequency hearing in smokers (Cunninham *et al.* 1983).

## SMOKING AND PREGNANCY

Cigarette smoking by pregnant women has been well known to exert deleterious effects on their infants. Fetal weight retardation has been recognized in almost all studies. The causal mechanism may be the direct effect of nicotine, fetal hypoxia produced by CO, or poor weight gain by smoking women. Exposure to 180 ppm (16-18% COHb) during pregnancy has been shown to result in 20% decrease of birth weight and a neonatal mortality of 35% as against 1% in the control group (Astrup *et al.* 1972). The relation between the number of cigarettes smoked by the pregnant mother and the low birth weight of the newborn is shown in figure 4-1.

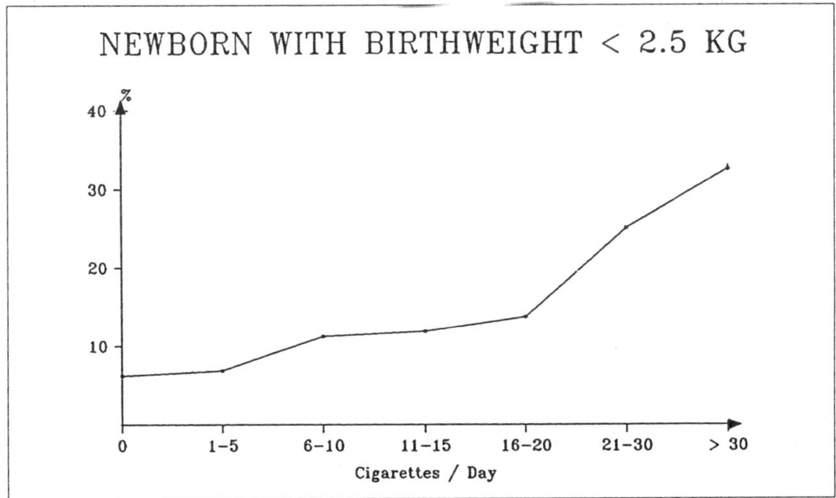


Fig. 4-1: Percentage of newborn with weight under 2.5 kg in relation to the number of cigarettes smoked by the mother during pregnancy (based on the data of Simpson 1957).

Tachi and Aoyama (1983) studied the effect of cigarette smoke and CO inhalation on the conceptus weight in pregnant rats; both lowered the conceptus weight on the days 14th and 21st of pregnancy. The effects were more pronounced in the group exposed to cigarette smoke as compared to the group exposed to CO.

This study suggests that CO is an important but not the sole factor responsible for the adverse effects of cigarette smoke on the fetus. This effect was most marked during the last third of pregnancy which is the phase of rapid fetal growth.

A significant effect of paternal smoking on birth weight has recently been shown with control for the effect of maternal smoking (Rubin *et al.* 1986, Schwartz-Bickenbach *et al.* 1987, Hadow *et al.* 1988).

Passive smoking during pregnancy has been found to double a non-smoker's risk of having a low birth weight infant (Martin and Bracken 1986, Nakamura *et al.* 1987). Collins *et al.* (1985) reported that pregnant rats exposed to smoke daily from day 5 to day 20 of gestation, when compared to control rats, showed reduced lung volume at term and saccules that were reduced in number. The internal surface area of the lung was decreased.

In non-smoking pregnant diabetic women in the third trimester, erythrocyte 2, 3-DPG has been found to increase as compared to healthy women, probably as an adaptive response to excess amounts of Hb A/C. In a study of smoker diabetic pregnant women, erythrocyte 2, 3-DPG was found to be lower than in the non-smokers (Madsen and Ditzel 1984). Arterial oxygen saturation and $PaO_2$ were reduced in the same proportion in both the smoker as well as non-smoker diabetic pregnant women. This study suggests that the disturbances in oxygen release are more marked in the smoking diabetic pregnant diabetic women as compared to non-smoking ones. If hypoxia is considered to be the cause of intrauterine death in diabetes, pregnant diabetic women should be advised not to smoke.

## CONCLUDING REMARKS

Smoking is an important cause of chronic low grade CO poisoning. Prevention of CO poisoning includes control of smoking (see Chapter 11). Smokers are more liable to develop high levels of COHb on environmental exposure to CO than non-smokers. Any efforts to treat chronic CO poisoning without smoking cessation are futile. COHb monitoring should be included in an anti-smoking program. Trell *et al.* (1983) showed that COHb was lower by 0.5% in those who had quit smoking than those who did not quit. Those who continued to smoke up to 10 cigarettes a day showed a rise in the median value of COHb by 1.2%. Alveolar CO measurements correlate highly with COHb levels and can be considered as a monitoring technique in anti-smoking programs (Wald *et al.* 1981).

Random COHb level determinations in men aged 34-64 years show the following values: cigarette smokers 4.7%, cigar smokers 2.9%, pipe smokers 2.2% and non-smokers 0.9% (Wald 1981).

Non-smokers, when exposed to 225 ppm CO, have a greater increase in COHb (from less than 2% to over 9%) than smokers who show only an increase from 8% to 12% (Vogel 1972). This leads to the question: Do smokers handle CO differently than non-smokers? This question has not been answered as yet and requires further investigations to assess the adaptation of smokers to CO.

Blood donors with high COHb values due to smoking pose a risk particularly when the recipients are infants. COHb values of 1.3% in an infant rise to 4.9% after transfusion of blood with COHb content of 5.2% (Kandall *et al.* 1973). In order not to eliminate smokers as blood donors, Shields (1971) recommended that they should abstain from smoking for 12 hours prior to blood withdrawal. Breathing of 100% oxygen prior to blood withdrawal has been suggested as an alternative.

## REFERENCES

Allen DR, Browse NL, Butt DL (1989) Effects of cigarette smoke, carbon monoxide and nicotine on the uptake of fibrinogen by the canine arterial wall.
Atherosclerosis 77:83-88.

Anderson G, Dalhamn T (1973) Health risks due to passive smoking (in Swedish).
Läkartidningen 70:2833-2836.

Aronow WS, Cassidy J, Vangrow JS et al (1974) Effect of cigarette smoking and breathing carbon monoxide on cardiovascular hemodynamics in anginal patients.
Circulation 50:340-347.

Astrup P, Trolle D, Olsen HM et al (1972) Effect of moderate carbon monoxide exposure on fetal development.
Lancet 2:1220-1222.

Astrup P (1973) Carbon monoxide, smoking and atherosclerosis.
Postgraduate Medical Journal 49:697.Badre RR, Guillerme N, Abram N et al (1978) Pollution atmospherique par la fumée tabac.
Ann Pharm Fr 36:443-452.

Benowitz NL, Jacob P III, Yu L et al (1986) Reduced tar, nicotine, and carbon monoxide exposure while smoking ultra low but not low yield cigarettes.
JAMA 256:241-246.

Benowitz NL (1989) Health and public policy implications of the low "yield" cigarette (editorial).
NEJM 320:1619-1621.

Bierenbaum ML, Fleischman AI, Steier A (1978) Effect of cigarette smoking upon in vivo platelet function in man.
Thromb Res 12:1051-1057.

Bogousslavsky J, Regli F, Van Melle G (1985) Risk factors and concomitants of internal carotid artery stenosis or occlusion: A controlled study of 159 cases.
Arch Neurol 259:864-867.

Burling TA, Stitzer ML, Bigelow GE et al (1985) Smoking topography and carbon monoxide levels in smokers. In,
Addictive Behaviors, Vol 10, New York, Pergamon Press, pp 319-323.

Butt J, Davies GM, Jones JG et al (1974) Carboxyhemoglobin levels in blast furnace workers. Ann Occupat Hyg 17:57-53.

Cano JP, Catalin J, Badre R et al (1970) Determination de la nicotine par chromatographie en phase gaseuze. II. Applications. Ann Pharmacol Fr 28:633-640.

Castleden CM, Cole PV (1975) Carboxyhemoglobin levels of smokers and non-smokers in the city of London. Brit J Indust Med 32:115-118.

Chappell SB, Parker RJ (1977) Smoking and carbon monoxide levels in enclosed public places in New Brunswick. Canad J Public Health 68:159-161.

Coburn RF, Foster RE, Kane PB (1965) Considerations on the physiological variables which determine the blood carboxyhemoglobin concentrations in man. J Clin Invest 44:1899-1910.

Collins MH, Moessinger AC, Kleinerman J et al (1985) Fetal lung hypoplasia associated with maternal smoking. Morphometric analysis. Pediat Res 19:408-412.

Crouse JR, Toole JF, McKinney WM et al (1987) Risk factors for external carotid artery atherosclerosis. Stroke 18:990-996.

Cuddleback JE, Donovan JR, Burg WR (1976) Occupational aspects of passive smoking. Am Ind Hyg Assoc J 37:263-267.

Cunningham DR, Vise LK, Jones LA (1983) Influence of cigarette smoke on extra-high frequency auditory thresholds. Ear and Hearing 4:162-165.

Dahms TE, Bolin JF, Slavin RG (1981) Passive smoking. Effects on bronchial asthma. Chest 80:530-534.

Davis JW, Davis RF (1979) Acute effects of tobacco cigarette smoking on the platelet aggregation ratio. Am J Med Sc 278:139-143.

Donnal GA, McNeil JJ, Adena MA et al (1989) Smoking as a risk factor for cerebral ischemia. Lancet ii:643-647.

Duncan DB, Greaney PP (1984) Passive smoking and uptake of carbon monoxide in flight attendants. JAMA 251:2658-2659.

Dwyer JH, Ndakorerwa GE, Semmer Nk et al (1988) Low level cigarette smoking and longitudual changes in serum cholestrol among adolescents. JAMA 259:2857-2862.

Effeney DJ (2987) Prostacyclin production by the heart: Effect of nicotine and carbon monoxide. J Vasc Surg 5:237-47.

Effenberger E (1957) Das Kohlenoxid und dessen Bedeutung in der Hygiene. Med Meteorol Hefte 12:1-128.

Fischer T, Weber A, Grandjean E (1978) Air pollution due to tobacco smoke in restaurants. Int Arch Occup Environ Health 41:267-280.

Fischer P, Richards JW (1986) The pathogenesis of atherosclerosis—and an exchange on research support by the tobacco industry (letter). NEJM 315:644.

Fielding JE, Phenow KJ (1988) Health effects of involuntary smoking. NEJM 319:1452-1460.

Ford CS, Crouse JR III, Howard G et al (1985) The role of plasma lipids in carotid bifurcation atherosclerosis. Ann Neurol 17:301-303.

Foliart D, Benowitz NL, Backer CE (1982) Passive absorption of nicotine in airline flight attendants. NEJM 308:1105.

Gelmini G, Dall'asta D, Palummeri E et al (1985) Fumo di sigaretta e deformabilita eritrocitaria. Gior Geront 33:321-322.

Godin G, Wright G, Shephard RJ (1972) Urban exposure to carbon monoxide. Arch Environ Health 25:305-313.

Goldsmith JR (1974) Carbon monoxide and smoking. Lancet 1:1232.

Grigani G, Gamba G, Ascari E (1977) Cigarette smoking effect on platelet function. Thromb Hemostasis 37:423-428.

Grünwald F, Schrock H, Kuschinsky W (1988) The effect of an actue infusion on local glucose utilization of the awake rat. Klin Wochenschr 66 (Suppl. XI):37-41.

Gvozdjakova A, Bada V, Sany L et al (1984) Smoke cardiomyopathy: disturbance of oxidative processes in myocardial mitochondria. Cardiovascular Res 18:229-232.

Hadow JE, Knight GJ, Palomaki GE et al (1988) Second trimester serum coninine levels in non-smokers in relation to birth weight. Am J Obstet Gynecol 159:481-484.

Hammond EC, Horn D (1958) Smoking and death rates - report on 44 months of follow up of 187783 men. JAMA 166:1159-1172 and 1294-1308.

Harke HP (1970) Zum problem des "Passivrauchens".
Münch Med Wochenschr 51:2328-2334.

Harke HP, Bleichert A (1972) Zum Problem des Passivrauchens.
Int Arch Arbeitsmed 29:312-322.

Harke HP, Peters H (1974) Zum Problem des Passivrauchens. III. Über den Einfluss des Rauchens auf die CO-Konzentration im Kraftfahrzeug bei Fahrten im Stadtgebiet.
Int Arch Arbeitsmed 33:221-229.

Harmsen H, Effenberger E (1957) Tabakrauch in Verkehrsmitteln, Wohn- und Arbeitsraümen.
Arch Hyg Bakteriol 141:383-400.

Hartz AJ, Anderson AJ, Brooks HL et al (1984) The association of smoking with cardiomyopathy.
New Engl J Med 311:1202-1206.

Haynes RL (1983) CO poisoning from non-tobacco cigarettes.
J Med Assoc Ga 72:553-555.

Hecht E, Vogt TM (1985) Marijuana smoking: effect on expired air carbon monoxide levels.
The International Journal of Addictions 20: 353-361.

Hoegg UR (1984) Cigarette smoke in closed spaces.
Environ Health Perspect 2:117.

Huch RJ, Danko L, Spatling L et al (1980) Risks the passive smoker runs.
Lancet 2:1376.

Hugod C, Hawkins LH, Astrup P (1978) Exposure of passive smokers to tobacco smoke constituents.
Int Arch Occup Environment health 42:21-29.

Jain KK (1990) Cerebral Insufficiency, Chicago, Year Book Medical Publishers.

Jarvis MJ, Russell MAH, Feyerabend C (1983) Absorption of nicotine and carbon monoxide from passive smoking under natural conditions of exposure.
Thorax 38:829-833.

Jarvis MJ, Russell MAH (1984) Measurement and estimation of smoke dosage to non-smokers from environmental tobacco smoke.
Europ J Res Dis 65 (Suppl. 133) 68-75.

Kachulis CT (1981) Second hand cigarette smoke as a cause of chronic CO poisoning.
Postgraduate Med 70:77-19.

Kandall SR, Landaw SA, Thaler MM (1973) Carboxyhemoglobin exchange between donors and recipients of blood transfusions.
Pediatrics 52:716-718.

Kanzler M, Jaffe JH, Nee J (1983) Low nicotine cigarettes; cigarette consumption and breath CO after one year.
Clin Pharmacol Ther 34:408-415.

Klausen K, Andersen C, Nandrup S (1983) Acute effects of cigarette smoking and inhalation of carbon monoxide during maximal exercise.
Europ J Appl Physiol 51:371-379.

Lawther PJ, Commins BT (1970) Cigarette smoking and exposure to carbon monoxide.
Ann N Y Acad Sc 174:135-147.

Levine PH (1973) Smoking and platelet function.
Circulation 48:619-623.

Levy AL, Lum G, Abeles FJ (1976) Carbon monoxide in firemen before and after exposure to smoke.
Ann Clin Lab Sc 6:455-458.

Lightfoot NF (1972) Chronic carbon monoxide exposure.
Proc Roy Soc Med 65:798-799.

Madsen H, Ditzel J (1984) Effect of smoking on red cell oxygen transport and release in diabetic pregnancy.
Act Obstet Gynecol Scand 63:77-80.

Martin TR, Bracken MB (1986) Association of low birth weight with passive smoke exposure in pregnancy.
Am J Epidemiol 124:633-642.

Mulchay R, McGilvray JW, Hickey N (1970) Cigarette smoking related to geographic variations in coronary heart disease mortality and to expectation of life.
Amer J Pub Health 60:1515-1521.

Nakamura M, Oshima A, Hiyama T et al (1988) effect of passive smoking during pregnancy on birth weight and gestation; a population-based prospective study in Japan.
In: Aoki M et al (eds): Smoking and Health 1987: proceedings of the sixth World Conference on Smoking and health, November 9-12, 1987, New York, Elsivier, pp 267-269.

National Research Council, Committee on Passive smoking (1986) Environmental Tobacco Smoke, Washington, DC, National Academy Press.

Nemeth-Coslett R, Griffiths RR (1985) Effects of cigarette rod length on puff volume and carbon monoxide delivery in cigarette smokers.
Drug and Alcohol Dependence 15:1-13.

Nil R, Woodson PP, Bättig K (1986) smoking behavior and personality patterns of smokers with low and high CO absorption.
Clincial Science 71:595-603.

Pankow D, Pankow B, Ponsold W (1976) gibt es bei Rauchern Anpassungsreaktionen an die Verminderung der Sauerstofftransportkapazität des Blutes infolge der chronischen Kohlenmonoxid-Inhalation?
Deutsch Gesundheit-Wesen 31:2045-2048.

Perry J (1973) Fasten your seat belts.
British Columbia Medical Journal 15:304-305.

Passero S, Rossi G, Nardinia M et al (1987) Italian Multicenter Study of reversible cerebral ischemic attacks, part 5: Risk factors and cerebral atherosclerosis.
Atherosclerosis 63:211-224.

Pimm PE, Silverman F, Shephard RJ (1978) Physiological effects of acute passive exposure to cigarette smoke.
Arch Environ helath 33:210-213.

Polak E (1977) Le papier a cigarette. Son role dans la pollution des lieux habites. Tabagisme passif: Notion nouvellee precise.
Brux Med 57:335-340.

Porthein F (1971) Zur Problem des "Passivrauchens".
Münch Med Wochenschr 118:707-709.

Palmer JR, Rosenberg L, Shapiro S (1989) "Low yield cigarettes and the risk of non-fatal myocardial infarction in women.
NEJM 320:1569-1573.

Pfueller SL, Burns P, Mak K et al (1988) effects of nicotine on platelet function.
Hemostasis 18:163-169.

Reynaud S, Blanche D, Dumont E (1984) Platelet function after cigarette smoking in relation to nicotine and carbon monoxide.
Clin Pharmacol Ther 36:389-395.

Rogers RL, Meyer JS, Judd BW et al (1985) Abstention from cigarette amoking improves cerebral perfusion among elderly smokers.
JAMA253:2970-2974.

Rubin DH, Krasilnikoff PA, Leventhal JM et al (1986) Effect of passive smoking on birth weight.
Lancet 2:415-417.

Russell MAH, Cole PV, Brown E (1973) Absorption by non-smokers of carbon monoxide from room air polluted by tobacco smoke.
Lancet 1:576-579.

Sarma JSM, Tillman SH, Ikeda S et al (1975) the effect of carbon monoxide on lipid metabolism of human coronary arteries.
Atherosclerosis 22:193-198.

Sebben J, Pimm P, Shephard RJ (1977) cigarette smoke in enclosed public facilities.
Arch Environ Health 32:53-58.

Seiff HE (1973) Carbon monoxide as an indicator of cigarette-caused pollution levels in intercity buses.
Publ. No. BMCS-IHA-73-1. Washington, D.C.:U.S. Department of Transportation, Bureau of Motor Carrier Safety, p. 13.

Schmidt JG, Rasmussen JW (1982) Acute platelet activation induced by smoking.
Thromb Hemostasis 51:279-282.

Schwartz-Bickenbach D, Schulte-Hobein B, Abt S et al (1987) Smoking and passive smoking during pregnancy and early infancy: effects on birth weight, lactation period, and cotinine concentrations in mother's milk and infant's urine.
Toxicol Lett 35:73-81.

Slavin RG, Hertz M (1975) Indoor air pollution: A study of the thirtieth Annual Meeting of the American Academy of Allergy, San Diego, Feb. 15-19.

Seppänen A, Uusital AJ (1977) Carboxyhemoglobin saturation in relation to smoking and various occupational conditions.
Ann Clin Res 9:261-268.

Shields CE (1971) Elevated carbon monoxide levels from smoking in blood donors.
Transfusion 11:89-93.

Simpson WJ (1957) a preliminary report on cigarette smoking and the incidence of prematurity.
Am J Obstst Gynecol 73:808-815.

Smith JR, Landau SA (1978) Smoker's polycythemia.
NEJM 298:6-10.

Srch M (1967) On the significance of carbon monoxide in cigarette smoking in an automobile.
Dtsch Z Gesamte Gerichtl 60:80-89.

Strong JP, Richards ML (1976) Cigarette smoking and atherosclerosis in autopsied men.
Atherosclerosis 23:451-476.

Surgeon General (1988) The Health consequences of Smoking: Nicotine Addiction
Washington, D.C., Government printing Office

Szadowski D, Harke HP, Angerer J (1976) Kohnlenmovoxidebelastung durch Passivrauchen in Büroraumen.
Zentralbl Prak Inn Med 3:310-313.

Tachi N, Aoyama M (1983) effect of cigarette smoke and carbon monoxide inhalation by gravid rats on the conceptus weight.
Bull Environ Contam Toxicol 31:85-92.

Tell GS, Howard G, McKinney WM et al (1989) cigarette smoking cessation and extracranial carotid atherosclerosis.
JAMA 261:1178-1180.

Trell E, Trell L, Janz L et al (1983) Verbesserung der Motivation zur Rauchentwöhnung bei Individuen mit genetischen Risikofaktoren für Lungenkrebsentwickelung.
Prax Klin Pneumol 37 (suppl 1) 687-690.

Toping DL (1977) Metabolic effects of carbon monoxide in relation to atherogenesis.
Atherosclerosis 26:129-137.

U.S. Department of Transportation and U.S. Department of Health, Education, and Welfare. Health Aspects of Smoking in Transport aircraft. U.S. Department of Transportation, Federal Aviation Administration, and U.S. Department of health, Education, and Welfare, Washington, D.C.:NIOSH, 1971.85 pp.

Vogel JA, Gleser MA, Wheeler RC et al (1972) Carbon monoxide and physical work capacity. Arch Environment Health 24:198-203.

Wald N, Howard S, Smith PG et al (1973) Association between atherosclerotic disease and carboxyhemoglobin levels in tobacco smokers.
Brit Med J 1:761-765.

Wald NJ (1976) Carbon monoxide as an etiological agent in arterial disease - some human evidence. In: Wynder EL, Hoffman D, Gori GB (eds). Proceedings of the 3rd World Conference on Smoking and Health, New York City, June 2-5, Vol. 1. Washington, D.C.: Government Printing Office (DHEW Publ. No. 76-1221), pp 349-359.

Wald NJ, Idle M, Boreham J et al (1981) Carbon monoxide in breath in relation to smoking and carboxyhemoglobin levels.
Thorax 36:366-369.

Weber A (1984) Annoyance and irritation by passive smoking.
Prev Med 13:618-625.

Weber AC, Jermini C, Grandjean E (1976) Irritating effects on man of air pollution due to cigarette smoke.
Am J Public Health 66:672-676.

Wolf PA, d'Agostino PB, Kannel WB et al (1988) Cigarette smoking as a risk factor for stroke. JAMA 259:1025-1029.

Yamashita K, Kobayashi S, Yamaguchi S et al (1988) Effect of smoking on regional blood flow in the normal aged volunteer.
Gerontolgoy 34:199-204.

Zacny JP, Stitzer ML, Yingling JE (1986) cigarette filter vent blocking: effects on smoking topography and carbon monoxide exposure.
Pharmacol Biochem Behavior 25:1245-1252.

# Chapter 5

## PATHOPHYSIOLOGY OF CARBON MONOXIDE POISONING

### - EFFECT ON VARIOUS SYSTEMS OF THE BODY -

### INTRODUCTION

Combination of carbon monoxide with hemoglobin to form COHb as well as the toxic effects of CO at cellular level have already been discussed in Chapter 2. These are the basis of pathological effects of CO on the human body. CO dissolved in the blood is more significant in the causation of tissue toxicity than the absolute COHb levels (Myers 1979). This is demonstrated by the metabolic damage caused by CO in an isolated brain perfused with HB-free blood.

Goldbaum *et al.* (1975) found that:

1. Dogs who inhaled 13% CO in air died in 15-60 min. CO levels at death were 54%-90% saturation.
2. Dogs rendered anemic by bleeding (Hb reduced by 68%) and then infused with Ringer solution and Dextran to restore volume survived after CO exposure even though the oxygen carrying capacity was impaired.
3. Dogs who were bled but infused with canine RBC containing 80% saturation of OHb showed a COHb level of 57&-64% saturation but showed no toxic symptoms.

Acute mortality from CO may be caused mainly by cerebral hypoxia and mediated mainly by cardiac dysrhythmias. Those who survive and show delayed neurological deterioration appear to have a combined hypoxic and ischemic insult during the acute exposure. Delayed neurological deterioration of CO poisoning has been thought to be mediated through lipid peroxidation and an example of the so-called ischemia-reperfusion injury (Marklund 1985). The biochemical changes occur only after CO poisoning, at a time when the COHb level is decreasing and thus probably, when there is improved oxygen delivery (Thom 1988).

Sub-lethal CO poisoning may result in some loss of glutamic acid decarboxylase activity and hence reduced ability to synthesize gamma amino butyric acid. This may contribute to the delayed neuropsychiatric sequelae of CO poisoning (Cheetam *et al.* 1988).

It was shown by Sokol (1984) in rat experiments that acute CO intoxication induced by long duration exposure did not involve CO accumulation in the tissues. Sokol (1985) also compared the effect of long versus short duration of exposure to CO in patients with similar COHb levels (40%) and showed that differences in biochemical effects could be detected by measurement of COHb.

There is considerable animal experimental data on CO poisoning but these findings cannot always be used to predict the effects of CO on humans. Lewey and

Drabkin (1944) demonstrated that chronic CO intoxication may occur in dogs at CO concentrations which have been regarded as being within the safety limits for man. The alleged sensitivity of canaries to CO is not due to a higher Haldane's coefficient of canary blood for CO but rather due to their more rapid rate of breathing. Their higher aerobic metabolism allows the canaries to reach equilibrium between blood and inspired CO more rapidly than humans (Spencer 1962).

TABLE 5-1
EFFECT OF CO ON VARIOUS SYSTEMS OF THE BODY

**Cardiovascular system**

- Precipitation of myocardial ischemia in patients with angina
- ECG abnormalities
- Cardiomyopathy as an acute effect and cardiomegaly as a chronic effect
- Hypertension and atherosclerosis as chronic effects

**Elements of the blood and hemorheology**

- Increased platelet aggregation
- Decreased RBC deformability
- Increased plasma viscosity and hematocrit
- Erythrocytosis as a chronic effect

**Nervous system**

- Brain: cerebral edema, focal necrosis
- Peripheral nerves: neuropathy and delayed motor conduction velocity

**Special senses**

- Visual system: retinopathy and visual impairment
- Auditory system: hearing loss due to hypoxia of the cochlear nerve

**Lungs:** pulmonary edema

**Muscles:** myonecrosis, compartment syndrome

**Exercise physiology:** decrease of physical work capacity and VO2 max

**Liver:** impaired function due to inhibition of cytochrome P-450

**Kidneys:** impairment of renal function, renal shutdown

**Endocrines:** impairment of hypophysis, hypothalamus and suprarenals

**Bone and Joints:** degenerative changes, hypertrophy of bone marrow

**Skin:** erythema and blisters

**Reproductive system**

- Impaired menstruation and fertility in women
- Impotence in men
- Fetal toxicity with low conceptus weight and growth retardation

The effects of CO on the human body are listed in Table 5-1 according to the system which is involved mainly even though the effects are not confined to any one system. Some animal experimental studies are cited. The pathology of lesions caused by CO is described in Chapter 6 and the clinical effects are discussed in more detail in Chapter 7.

## CARDIOVASCULAR SYSTEM

### Acute effects on the heart

The heart is particularly vulnerable to CO poisoning because CO binds to the cardiac muscle three times as much as to the skeletal muscle. Studies on isolated animal hearts have shown that CO may have a direct toxic effect on the heart regardless of the formation of COHb (Chen and McGrath 1985). At levels of 1%-4% COHb, myocardial blood flow is higher but no adverse effects are demonstrated. If the perfusion medium of an isolated rat heart muscle is gassed with 10% CO, there is 40% increase in coronary blood flow, which is likely to be due to vasodilatation secondary to anoxia (McGrath 1984). Increase in myocardial blood flow occurs mostly without an increase in the COHb levels. Vasodilation has been explained due to CO's interfering with $CA^{++}$ flux in vascular smooth muscle (Lin and McGrath 1988).

Melinshyn *et al.* (1988) showed that a compensatory rise in cardiac output occurs during CO hypoxia in dogs when the level of COHb exceeds 40%. They concluded that this depends on peripheral vasodilaiton mediated through β-adrenoreceptors.

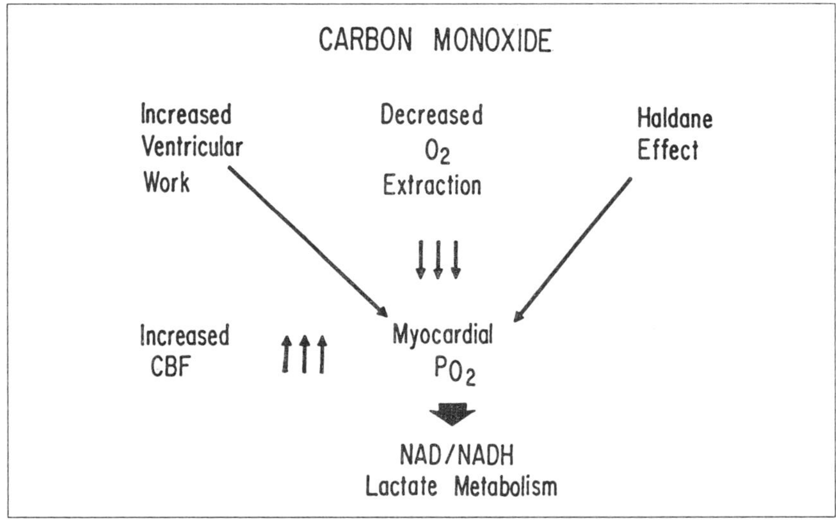


Fig. 5-1: Mechanism of decrease of myocardial oxygen tension by carbon monoxide *(Reproduced from Ayers et al. 1970, by permission)*. See text for explanation.

Ayers *et al.* (1970) have pointed out that the level of myocardial oxygenation is a function of both oxygen demand and oxygen need (Fig. 5-1). CO decreases the myocardial oxygen tension by three mechanisms:

1. Increased ventricular work and oxygen demand due to stimulation of the adrenergic system.

2. Decreased oxygen extraction.

3. Decreased capillary oxygen tension because of leftward shift of the oxyhemoglobin dissociation curve. This illustration further shows that the major adaptive defense to this combined assault is an increase in coronary blood flow. Measurement of myocardial lactate and pyruvate metabolism serve as indicators of the oxidative metabolism in the myocardial cell and may be used to determine the net effect of the factors listed above.

Nag *et al.* (1988) cultured embryonic rat cardiac cells in presence of various tensions of CO (5-95%) in the absence of oxygen and demonstrated reduced cell growth which was concentration dependent. When cardiac cells were grown in the presence of a mixture of CO (10-20%) and oxygen (10-20%), the growth rate of these cells was comparable to those of cells under normal atmospheric conditions. There was no significant difference in the ultrastructural organization of mitochondria in the latter experiment. The authors pointed out that the alterations in the cells may occur at enzymatic level and not reflect the structural changes.

The effects of exposure to CO on the human cardiovascular system are shown in Table 5-2.

TABLE 5-2
EFFECT OF CO EXPOSURE ON THE HUMAN CARDIOVASCULAR SYSTEM

[After, Rylander R, Vesterlund J: Carbon monoxide criteria. With reference to effects on the heart, central nervous system and fetus. Scan J Work Environ Health 1981; (suppl 1):17, by permission]

| CO AND/OR COHb LEVEL | DURATION | EFFECTS |
|---|---|---|
| 5% CO<br>9% COHb | 30–120 s | Increase in coronary blood flow and oxygen extraction ratio and decrease in coronary sinus oxygen tension in subjects with normal hearts; no significant increase in coronary heart disease subjects at high COHb levels; oxygen extraction ratio increased and coronary sinus tension decreased in heart patients, altered lactate and pyruvate metabolism seen at high COHb levels |
| 100 ppm<br>4.08% COHb | 1 h | Significant decrease in exercise time until marked dyspnea in subjects with chronic obstructive pulmonary disease |
| 42–63 ppm<br>5.08% COHb | 90 min | Exercise time until onset of angina decreased significantly in angina pectoris patients; systolic blood pressure and heart rate at angina also significantly decreased |
| 50 and 100 ppm<br>2.9 and 4.5% COHb | 4 h | Exercise time until onset of angina in angina pectoris patients reduced even at low COHb level; systolic blood pressure and heart rate also decreased at angina |
| 7 and 20% COHb | | Maximal exercise time for healthy subjects decreased with increasing COHb levels; no difference in blood lactate concentrations between exposed and control groups; maximum oxygen uptake decreased with increased COHb |
| 100 ppm<br>3.95% COHb | 1 h | Mean exercise time until exhaustion decreased in middle-aged, healthy nonsmokers; reduced work time for exposed nonsmokers; no reduction in exposed smokers; respiratory pattern changes noted in smokers |
| 50 ppm | | Study of "admission case fatality rate" in hospitals divided into high- and low-polution areas: significant difference found for fatality rates between high- and low-pollution area hospitals in 9 of 10 weeks when ambient CO levels were above a weekly mean of 9 ppm |
| | | Significant low-level association observed between ambient CO levels and frequency of initial cardiorespiratory complaints at an emergency room of a hospital in Denver, Colorado, during high-pollution months |

Patchy myocardial necrosis with characteristic changes of ischemia is seen in rabbits exposed to CO levels of 180 ppm for 2 weeks (Wenstrup *et al.* 1969). Left ventricular wall motion abnormalities, as shown by echocardiography are frequently seen in CO poisoning and these correlate with a high incidence of papillary muscle lesions in fatal cases (Corya *et al.* 1976).

Angina patients are particularly susceptible to CO exposure. Onset of angina during physical exertion can be accelerated by elevating COHb levels to the 5-9% range. Studies reporting the effect of CO on angina pectoris are summarized in Table 5-3. Dwyer and Torino (1989) have reviewed this subject recently.

TABLE 5-3
STUDIES OF THE EFFECT OF CO EXPOSURE ON ANGINA PECTORIS

| *Author and year* | *Study* | *Methods* | *Results* |
|---|---|---|---|
| Aronow 1979a | Effect of 4% COHb on human performance in cardiac patients. | Double-blinded crossover study. 20 men with angina pectoris. | Significant impairment of visualization test after breathing CO. |
| Aronow 1979a | Effect of non-nicotine cigarettes and CO on angina pectoris. | Double-blinded crossover, 12 men. COHb to 5.35%. | Decreased exercise time to angina. ST-segment depression. Smokers worse. |
| Aronow 1981 | Aggravation of angina pectoris by 2% COHb. | Double-blinded randomized, crossover study. 15 men breathing 50 ppm CO for 1h. | Exercise duration to angina reduced. Decrease in systolic BP and heart rate at time of angina. |
| Koskela *et al.* 1976 | Angina pectoris, ECG findings and BP of foundry workers in relation to CO exposure. | Prevalence study in foundry in relation to CO exposure. | Higher prevalence of possible angina history but not ECG changes with greater CO exposure. |
| Anderson *et al.* 1973 | Effect of CO on patients with stable angina pectoris. | Exposure to 50 ppm or 100 ppm CO on 5 successive days | Exposure to CO 100 ppm (but not 50 ppm) aggravated myocardial ischemia. |
| Allred *et al.* 1989 | 63 patients with myocardial ischemia due to coronary artery disease. | Multicenter study. Exercise to ischemic threshold (S-T segment depression). Exposure to CO (COHb level 1-4%) | Objective demonstration of decrease of 5.1% in the length of time to ischemic threshold after CO exposure. Even low levels of COHb can exacerbate myocardial ischemia in patients with coronary artery disease. |

A 90-minute exposure on Los Angeles Freeway has been shown to cause ECG abnormalities in 40% of the patients with pre-existing heart disease (Ayers *et al.* 1973). Atkins *et al.* (1985) reported the death of two patients with coronary heart disease after exposure to CO. They pointed out that CO precipitated ischemia by reducing oxygen delivery to the myocardium. Changes in ECG have been shown to occur in workers exposed chronically to CO when COHb levels reach 20-30% (Komatsu 1970). These changes are reversible after removal from CO exposure. Various ECG abnormalities reported in CO poisoning are listed in Table 5-4. Depression of ST segment is the most common ECG finding in these cases and may precede myocardial infarction.

TABLE 5-4
EEG ABNORMALITIES DUE TO CO POISONING

1. Arrythmias, extrasystoles, atrial fibrillation.
2. Low voltage
3. Depression of ST segment
4. Prolongation of ventricular complex, particularly the Q-T interval.
5. Conduction defects: increased P-R interval, A-V block, bundle branch block.

Conduction abnormalities may be due to anoxia or direct toxic effect of CO or hemorrhages into the conducting system of the heart. Ginsberg and Myers (1974) demonstrated that ECG abnormalities and ventricular arrythmias occurred in all, and ventricular fibrillation in 25% of the monkeys gassed with CO. Exposure of dogs to 100 ppm of CO for 2 h with rise to COHb to 6.48% has been shown to decrease ventricular fibrillation threshold.

Recent studies have shown that acute exposure to CO resulting in COHb levels of 4-6% does not increase the frequency or severity of ventricular atopy in patients without significant arrythmias at baseline (Hinderliter *et al.* 1989).

The practical implication of various studies of the effect of CO on the cardiovascular system is that low levels of COHb (2%) may be tolerated by healthy individuals but those with cardiovascular disease are at risk. It is estimated that about 10 million of the population of United States is exposed to CO levels which result in COHb levels of 2% or more (Kuller and Radford 1983). Considering that the prevalence of coronary heart disease in adults is about 20/1000, 200,000 such patients are exposed to the detrimental effects of environmental CO. About 25% of these patients (50,000) have major coronary artery disease which increases the risk of heart attack and sudden death. If this risk is 5% per year and CO exposure doubles this risk, 5000 deaths may be expected annually due to low dosage environmental CO exposure. At present there is no way to verify this supposition.

## Chronic effects on the cardiovascular system

Impaired oxygen delivery to the arterial wall tissue has been proposed as a factor in the etiology of atherosclerosis. The role of CO-induced arterial wall hypoxia has been reviewed by Schneiderman and Goldstick (1978), who used a computer simulation of the oxygen supply system of the arterial wall. Their conclusion supports the concept that mid to inner wall hypoxia, coupled with abnormal lipoprotein infilteration resulting from CO-induced endothelial injury, may be the two major interacting factors in the atherogenic process.

Cardiomegaly has been shown in adult male rats inhaling 500 ppm CO continuously for 38-47 days (Penney *et al.* 1984). They found that the left ventricular wall was slightly thicker than the interventricular septum, while right ventricular wall thickness was about 35% that of left ventricular wall thickness. These findings suggest that chronic carboxyhemoglobinemia produces an eccentric cardiomegaly with no intrinsic changes in wall stiffness. Evidence of myocarditis was found in 18% of Japanese farmers exposed to a mean CO concentration of 70 ppm (Komatsu 1959).

Exposure to CO has been shown to cause hypertension in Dahl rats (a strain which is usually susceptible to hypertension) who were fed a high sodium diet. Controls fed the same diet did not develop hypertension (Shiotsuka *et al.* 1984). The results of this study cannot be transposed to humans as the COHb levels in rat experiments were 42%, which would be severely incapacitating in humans for whom a 67% COHb level is considered to be lethal.

Chronic CO exposure is a risk factor for cardiovascular disease by increasing atherosclerosis. Astrup *et al.* (1970) carried out a series of experiments in cholestrol fed rabbits exposed to various concentrations of CO. In animals exposed to CO for 10 weeks (15% COHb for 8 weeks, 30% the last 2 weeks), the cholesterol content of aorta was 2.5 times that of control animals not exposed to CO but fed cholestrol. The cholestrol content rose to 5 times in animals exposed daily for 8 hours to CO concentrations leading to 20% COHb over a period of 10 weeks. If the content of oxygen in the breathed air was raised to 28%, the content of cholesterol dropped to half the level in control animals. There was no difference in the arterial lesions produced by exposure to CO and those produced by hypoxia without CO.

Increased incidence of cardiovascular disease in chronic heavy smokers has already been described in Chapter 4. Heart disease mortality has been shown to be higher in bridge and tunnel workers exposed to CO (Stern *et al.* 1988). Experimental studies of chronic CO exposure in dog model of myocardial ischemia, however, do not support the clinical observations described (Vanoli *et al.* 1989).

## HEMATOLOGICAL AND HEMORHEOLOGICAL EFFECTS OF CO EXPOSURE

CO concentrations of 200 ppm in the inhaled air, which lead to COHb levels similar to those of moderate smokers, increase platelet aggregability. Exposure of experimental animals to high levels of CO has been shown to elicit rapid compensatory polycythemia and rise of hematocrit. Increase of the RBC volume and decrease in plasma volume in "smoker's polycythemia" has been attributed to CO (Smith and Landau 1978).

Chen *et al.* (1984) exposed Sprague-Dawley rats to 450 ppm of CO for six hours per day, five days per week, for 33 days. They found a three-fold increase in young reticulocyte population along with an increase in the total reticulocyte count. Despite continuous exposure, reticulocyte and distribution returned to normal by day four. Both hematocrit and Hb concentrations began to increase 16 days after CO exposure and remained elevated for the duration of the exposure. These authors also showed that CO-induced polycythemia is associated with cardiac hypertrophy in rats. Presumably, the CO-induced hypoxia evokes an increase of erythropoietin secretion in the kidneys and a stimulation of erythropoiesis (Pankow and Ponsold 1981a). This view is supported by the demonstration that 56% CIHb causes an increase delta aminolevulinic acid hydratase activity in the rat RBC even after a single exposure to CO (Pankow 1984). Concomitant exposure to other toxins with nephrotoxicity inhibits the polycythemic response to CO (Pankow and Ponsold 1981b). There are seasonal variations in the CO-induced polycythemia in rats (Pankow *et al.* 1983).

Increase of COHb levels also decreases the erythrocyte deformability, thus impairing the microcirculation.

Exposure to 400 ppm of CO can cause increase of viscocity of plasma.

## CHANGES IN BLOOD CHEMISTRY

Levels of COHb over 5% have been shown to raise blood lactate levels (Ayers *et al.* 1970). This is presumed to be an effect of hypoxia. Sokal *et al.* (1985) observed that the severity of CO poisoning depends on the duration of exposure rather than on COHb levels alone. Severe CO poisoning associated with long exposure is accompanied by high blood lactate and pyruvate.

Ceruplasmin activity in rat serum has been shown to be markedly inhibited by acute (58+5% COHb) or chronic ($36 \neq 10$ and $50 \neq 6$% COHb) carbon monoxide poisoning. This CO effect did not appear to be due to decreased $O_2$ transport capacity of the hemoglobin (Müller and Pankow 1978).

## THE NERVOUS SYSTEM

### The Brain

This is the organ most susceptible to the toxic effects of CO. The effects of anoxia are most marked in the brain because it consumes the largest proportion of oxygen of the whole organism. The adult brain consumes about a quarter of the oxygen inhaled; in children up to the age of 4 years, this proportion rises to a third.

CO produced anatomic changes in the brain which are described in Chapter 6. The classical explanation of cerebral edema induced by CO is tissue hypoxia but recent evidence indicates that severe CO intoxication may induce cerebral edema even in the absence of tissue hypoxia. The cytotoxic effect is mediated presumably via direct binding of CO to intracellular copper or heme containing proteins. Ligation of common carotid artery and exposure to CO is a reliable method of producing unilateral cerebral infarcts in rats (Laas *et al.* 1989). CO-induced hypoxia in these experiments, in contrast to hypoxic hypoxia, did not stimulate chemoreceptors, thus causing a severe systemic hypotension due to peripheral vascular dilation. An inhibition of ferro-enzymes by CO did not seem to be involved in these cases and the cause of death was shown to be diffuse ipsilateral cerebral edema. It is not certain if patients with cerebrovascular disease are more susceptible to the effects of CO poisoning.

### Effect on the peripheral nerves

There are several reports of the impairment of peripheral nerve function after CO intoxication. CO poisoning may produce:

1. Peripheral neuropathy
2. Reduction of amplitude of action potential of the isolated nerves.
3. Retardation of nerve conduction. Carboxyhemoglobinemia of 20% has been shown to decrease nerve conduction velocity, COHb of 50% shows no further alterations and further of COHb again leads to decrease of velocity in experimental animals (Pankow *et al.* 1975).

Possible causes of peripheral neuropathy in human CO poisoning are: hypoxia, toxic effect of CO on the nerves, and positional compression of peripheral nerves during the comatose state (Snyder 1970).

### Effects on the visual system

Measurable decreases in light sensitivity and adaptation to darkness have been shown to result from low level CO exposure. These alterations persist even after elimination of COHb from the blood, indicating that a significant amount of CO may be retained in the tissues. Retinal hemorrhages have been observed on ophthalmoscopy of patients with acute CO poisoning. Retinal venous engorgement and peripupillary hemorrhages resemble those seen in hypoxia. Carbon monoxide retinopathy has been recorded as an acute effect of CO poisoning and as a subacute effect (Kelly and Sophocleus 1975). Pathophysiological theories proposed to explain retinopathy of CO include Valsalva's manouver and compression of retinal venous drainage with cerebral edema. These changes are similar to those seen in high altitude mountain climbers and suggest hypoxia as a causal factor (Dempsey *et al.* 1976).

**Effect on the auditory system**

Hearing loss of a central type due to anoxia from CO poisoning is partially reversible. The loss of auditory threshold activity is pronounced at the auditory cortex level. Relative vulnerability of the central auditory pathway has been demonstrated (Makashima *et al.* 1977). Vestibular function is more frequently involved than the auditory function. Decrease in hearing loss is uncommon and is due to hypoxia of the cochlear nerve and the brainstem nuclei. Acute CO intoxication can potentiate noise-induced threshold shifts and increase hair cell loss associated with the functional loss (Fecther *et al.* 1988). Cochlear and brainstem hypoxia leads to central hearing loss and vestibular dysfunction (nausea, vomiting, vertigo), with vestibular symptoms usually more prominent than auditory loss (Baker and Lily 1977).

**Sleep**

Sleep is severely disturbed by CO without a detectable effect on the respiratory frequency and pulmonary ventilation. Lahiri *et al.* (1981) stated that the aortic body receptors mediating circulatory reflexes are more sensitive to CO than the carotid body receptors mediating respiratory reflexes. Disruption of sleep could result from afferent discharged from aortic receptors in response to CO or low oxygen content as anoxia is known to abolish REM sleep (Pappenheimer 1984).

## RESPIRATORY SYSTEM

The tissues of the respiratory system are exposed to the highest concentration of CO. Pulmonary edema is a prominent feature of the lung pathology and the various mechanisms of its production are: direct effect of CO on the capillary membranes with increase of permeability, direct effect on the lung parenchyma, indirect effect of cerebral edema and damage to the respiratory center, effect of hypoxia on the cardiovascular system.

Disturbances of the respiratory rhythm and impairment of the lung diffusion capacity are common in the survivors of CO poisoning.

**Effect on Skeletal Muscles**

King *et al.* (1987) investigated the consequences of a decreased oxygen supply to a contracting canine gastrocnemius preparation under two forms of hypoxia: hyposic hypoxia and CO-hypoxia. Oxygen uptake decreased with CO-hypoxia but not with hypoxic hypoxia. There was also a lesser oxygen extraction during

CO-hypoxia which the authors considered possibly due to leftward shift of the oxyhemoglobin dissociation curve and decreased myoglobin function.

Massive myonecrosis may occur as a result of CO poisoning, leading to edema, compartment syndrome, and acute renal failure (Finlay *et al.* 1977). The precise mechanism of this is not known. It may be related to cellular hypoxia or direct toxic effects. Shapiro *et al.* (1989) did not find any evidence of rhabdomylysis in their patients with CO poisoning as determined by creatine phosphokinase levels. They consider that cyanide in smoke inhalation injury may play a more important role in causing rhabdomyolysis.

## CO IN RELATION TO EXERCISE PHYSIOLOGY

The effect of hypoxia on physical exercise capacity has been discussed in detail elsewhere (Jain 1989).

Human studies of the physiological effects of CO began with the pioneer studies of Haldane on himself (Haldane 1895). He noticed no effect of acute CO exposure at rest on ventilation or pulse rate at COHb saturation less than 20%. When he walked home after the experiment, COHb saturation being still 18%, he became out of breath, and experienced dimness of vision, palpitations, and tinnitus.

Chiodi *et al.* (1941) showed that when the COHb levels rose to 45%, their subjects were unable to carry on tasks requiring levels of energy. Other investigators have also provided evidence of deterioration of aerobic power capacity as an effect of CO intoxication (Ekblom and Huot 1972, Klausen *et al.* 1968, Pirnay *et al.* 1968, Vogel *et al.* 1972). A linear decline in $VO_2$ max occurs with a progressive increase in COHb levels from 5-33% (Figure 5-2).

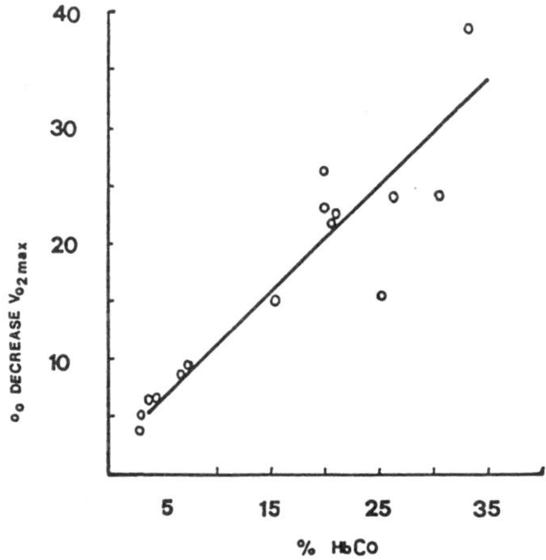


Figure 5-2: Relationship between carboxyhemoglobin levels and maximal aerobic capacity ($VO_2$max).

Seppänen (1977) studied the physical work capacity in relation to carbon monoxide inhalation and tobacco smoking. Physical work capacities at heart rates 130, 150 and 170 beats per minute decreased both after CO inhalation and after smoking (COHb 9.8%). Effect of smoking on exercise capacity has been discussed in Chapter 4.

Fatigue and reduced exercise capacity may also be due to accumulation of lactate resulting from CO exposure. Lactate levels over 4 m mol hinder physical training.

## GASTROINTESTINAL TRACT

Many of the gastrointestinal symptoms of CO poisoning such as nausea and vomiting are due to central nervous system involvement. Some of the effects are also due to hyperemia and hemorrhages of the mucous membranes.

## LIVER

Liver cells are also affected by CO-hypoxia. Zorn (1972) showed that in experimental CO poisoning in rats, there is a reciprocal linear relationship between $PO_2$ in the liver tissues and the COHb concentration in the blood. The $PvO_2$ in the portal veins in the rats drops after inhalation of 80 mg $CO/m^3$ and when it is increased to 1.2 g $CO/m^3$ the $PvO_2$ drops from 40 mm Hg to 10 mm Hg (Montgomery and Rubin 1973). This leads to disturbances of many of the liver functions. There is evidence that some effects of CO on the liver are independent of hemoglobin blockade:

1. Rat liver perfusion in situ with blood containing 10% $^{14}$COHb leads to retention of 0.30 µl $^{14}$ CO/g even though the liver is washed free of blood before the measurement (Topping 1975).

2. In animal experiment under variable conditions, CO exposure has been shown to lead to an inhibition of cytochrome oxidase activity in the liver.

3. In fatal cases of CO poisoning, a measurable blockage of cytochrome P-450 has been demonstrated in liver microsomes.

The pathology of liver lesions in CO poisoning is described in Chapter 6.

## KIDNEYS

The manifold tasks of the extraordinarily vascularized kidneys have high oxygen requirements. CO-hypoxia leads to various pathological changes in the kidneys (see Chapter 6). As a sequel of muscle necrosis or shock, oliguria or even renal shutdown may occur. Experimental studies in animals have shown that:

1. In rabbits exposed to CO (30% COHb) there is reduction of $PvO_2$ and renal oxygen utilization due to increase of renal oxygen extraction. There is decrease of sodium resorbtion as well (Bursaux *et al.* 1976).

2. CO intoxication in dogs leads to stimulation of release of erythropoietin which is formed in the kidneys (Syvertsen *et al.* 1973).

3. Decrease of renal blood flow and golmerular filteration after CO poisoning in rabbits.

4. Inhibition of diuresis after subcutaneous injection of CO in rats

5. Significant increase of urinary catecholamines after subcutaneous injection of CO but repeated injections lead to disappearance of this effect (Pankow and Ponsold 1978).

### Endocrine System

The effects of CO on the endocrine system are tied in with its effects on the nervous system. Changes in the function and morphology of the hypophysis, hypothalamus, and the suprarenals have been noted. Changes in the thyroid parynchyma and function have also been described after CO poisoning.

### Bone

It does not seem that bone is influenced directly by CO poisoning. Bour *et al.* (1966) stated that the degenerative joint disorders seen as sequelae of CO poisoning are more likely due to prolonged immobilization associated with comatose state. Anoxia, sympathetic stimulation, and vascular changes play an auxiliary role in bone changes in CO poisoning.

Bone marrow as an organ for formation of blood elements is of importance in the increased hematopioesis as a sequel of CO poisoning. Bone marrow of guinea pigs chronically exposed to CO is twice as large as the control animals not exposed to CO (Caccuri 1941).

### Skin

The classical cherry red color of the skin and the mucous membranes is characteristic of non-survivors, because the high COHb levels required to produce this appearance are usually not compatible with life.

Cutaneous blisters occur in CO poisoning (Myers *et al.* 1985). It seems possible that necrobiosis in eccrine glands starts early, but that the epidermal basal cells, notable at the papillary apices suffer the same change only after temporary pressure anoxia and reactive hyperemia. Findlay (1988) has made an optical study of the skin color in fatal cases of CO poisoning, using reflectance spectrophotometry; he pointed out several factors which contribute to the difficulty in identifying the classical cherry red color. These are: skin pigmentation, low CO concentration, and deep venous dilatation with superficial vasoconstriction producing the impression of cyanosis.

## EFFECTS ON THE REPRODUCTIVE SYSTEM

Effects on fertility and reproduction. Hypoxia is known to impair fertility. Exposure to CO produces menstrual disturbances in women and decrease of libido, erections, and potency in men. Rats exposed daily to CO inhalation for months show reduction of testicular weight and sperm count (Patterson *et al.* 1938). Repeated exposure of guinea pigs to CO disturbs the follicle ripening.

Effects on the fetus. Pregnant state enhances the effects of CO poisoning (see Chapter 2). Many authors have discussed the harmful effects of CO on fetal development (Wells 1933, Goldstein 1965, Fechter and Annau 1977). Direct poisoning of the fetus had been demonstrated when its carrier is exposed to CO. Fetal uptake of CO occurs 2 to 3 times slower than in the mother and is eliminated slowly from the fetus as well. Harm results not only from hypoxia but also from impairment of the placental metabolism.

Tachi *et al.* (1983) studied the effects of CO inhalation on the conceptus weight in gravid rats and reached the following conclusions:

1. Continuous CO inhalation lowered the conceptus weight on days 14 and 20 of pregnancy.
2. The effect was more pronounced in the group exposed to cigarette smoke (CO plus nicotine) than the group exposed to CO alone.
3. CO affects the fetus more adversely during the last trimester of pregnancy, which is the phase of rapid growth.

Experimental studies in neonatal animals have shown that acute CO exposure can alter neurotransmitter function in the brain and that some of the effects persist for several weeks (Newby *et al.* 1976). Exposure of neonatal rats to CO has also been shown to produce hyperactivity which persisted for up to three months of age (Culver and Norton 1976).

Cho and Yun (1988) exposed Sprague-Dawley rats to various concentrations of CO starting at day 11 of pregnancy when active organogenesis starts. Rats exposed to 1400 ppm of CO for 2 hours (COHb levels of 68-72%) sustained significant damage to the pregnancy. The fetuses of the rats treated by HBO 15 minutes after exposure to CO did not show any growth retardation. CO was not found to be tertogenic in any of the rats exposed to CO.

There is little epidemiological evidence regarding the effects of CO air pollution on birth weight. Mothers living in Los Angeles areas with high levels of air pollution have been reported to have delivered infants of substantially lower birth weight than mothers residing in areas with low levels of air pollution. Alderman *et al.* (1987) conducted a case control study of pregnant women in their last trimester who lived in Denver, Colorado which has reportedly high levels of environmental CO. They were unable to show any relationship between low birth weight and exposure to CO.

Other studies of the fetal effects of maternal carbon monoxide exposure associated with smoking are shown in Table 5-5.

TABLE 5-5

FETAL EFFECTS OF MATERNAL CARBON MONOXIDE EXPOSURE: STUDIES IN HUMANS
(Reproduced from: Rylander R, Vesterlund J: Carbon Monoxide criteria. Scand J Work Environ Health 1981:7:1-39).

| Investigators Year | Description | Effects |
|---|---|---|
| Astrup *et al.*, 1972 | smoking habits and COHb levels of 253 pregnant women recorded; birth-weights and conditions of neonates later recorded. | mean birthweight of neonates of non-smoking mothers was 3,225 g. and of smoking mothers 2,990 g |
| Goujard *et al.*, 1975 | prospective investigation of 6,989 pregnant women. | 250% increase observed in number of stillbirths among smoking mothers; a large proportion of the increase was due to abruptio placentae. |
| Meyer *et al.*, 1976 | further analysis of data collected in Ontario Perinatal Mortality Study. | neonatal birthweight and length of gestation seen to be directly related to mothers' smoking habits; placental complications increased with the level of smoking, except for mothers pregnant for the first time smoking less than 1 pack/day, particularly for placenta previa and abruptio placentae. |
| Fedrick *et al.*, 1971 | further analysis of data collected in 1958 British Perinatal Mortality Study and National Child Development Study, with reference to identifying possible connections between maternal smoking and incidence of congenital heart disease. | analysis demonstrated that the occurrence of congenital heart disease in babies was 50% greater among smoking than non-smoking mothers; sample too small, however, to allow for firm conclusions. |
| Buncher 1969 | data collected from obstetrical wards in Naval hospitals in U.S.; effects of smoking during pregnancy studied. | observed difference in length of pregnancy between non-smoking and smoking mothers (more than 1 pack/day) was 29 to 24 h; it was suggested that about 10% of the known reduction in birthweight could be attributed to this shortened gestation period. |

Caravati *et al.* (1988) reported 6 cases of acute CO poisoning during pregnancy. All the women survived with good outcome but in two cases the fetuses died. In a third case, the fetus was delivered at premature labor in the 33rd week of pregnancy. It had multiple morphological abnormalities and did not survive. The other three women had normal full term deliveries with healthy infants. These mothers were treated only with normobaric oxygen and not hyperbaric oxygen which might have improved the outlook.

# REFERENCES

Alderman BW, Baron AE, Savitz DA (1987) Maternal exposure to neighborhood carbon monoxide and risk of low infant birth weight.
Public Health Reports 102:410-414.

Allred EN, Bleecker ER, Chaitman BR et al (1989) Short-term effect of carbon monoxide exposures on the exercise performance of subjects with coronary heart disease. New Engl J Med 321:1426-1432.

Anderson EW, Andelman RJ, Strauch JM et al (1973) Effect of low-level carbon monoxide exposure on onset and duration of angina pectoris.
Annals of Int Med 79:46-50.

Aronow WS, Charter R, Seacat G (1979a) Effect of 4% carboxyhemoglobin on human performance in cardiac patients.
Prev Med 8:562-566

Aronow WS, Stemmer EA, Zweig S (1979b) Carbon monoxide and ventricular fibrillation threshold in normal dogs.
Arch Environ Health 74:184-186.

Aronow WS (1981) Aggravation of angina pectoris by two percent carboxyhemoglobin.
Am Heart 101:154.

Astrup P, Kjeldsen K, Wanstrup J (1970) Effects of carbon monoxide exposure on the arterial walls.
Ann NY Acad Sc 147:294-300.

Astrup P (1972) Some physiological and pathological effects of moderate carbon monoxide exposure.
Brit Med J 4:447-452.

Atkins EH, Baker EL (1985) Exacerbation of coronary artery disease by occupational carbon monoxide exposure: A report of two fatalities and a review of the literature.
Amer. J. Indust. Med. 7: 73-79.

Ayres M, Giannelli S, Mueller H (1973) Carboxyhemoglobin and the access to oxygen. An example of human counter-evolution.
Arch Environ Health 26: 8-15.

Ayres SM, Gianelli S, Mueller H (1970) Myocardial and systemic responses to carboxyhemoglobin.
Ann NY Acad Sc 170:268.

Baker Sr, Lilly DJ (1977) Hearing loss from acute carbon monoxide intoxication.
Ann Otorhinolaryngol 86:323-328.

Buncher CR (1969) Cigarette smoking and duration of pregnancy.
Am J Obstet Gynecol 103:942-946.

Bursaux E. Poyart C, Bohn B (1976) Renal hemodynamics and renal $O_2$ uptake during hypoxia in the anesthetized rabbit.
Pflügers Arch 365:213-220.

Caccuri S (1941) Sul consumo di ossigeno del middolo osseo e del fegato nell'intossicazioni acuta da ossido di carbonio.
Hematologica 23:165-175.

Caravati EM, Adams CJ, Joyce SM et al (1988) Fetal toxicity associated with maternal carbon monoxide poisoning.
Ann Emerg Med 17:714-717.

Cheetam SC, HJorton RW, Crompton MR et al (1988) Carbon monoxide poisoning.
Brit med J 296:787-788 (letter).

Chen KC, McGrath JJ (1985) Response of the Isolated Heart to Carbon Monoxide and Nitrogen Anoxia.
Toxicology and Applied Pharmacolgy 81:363-370.

Chen KC, Lee EW, McGrath JJ (1984) Effect of intermittent carbon monoxide inhalation on erythropoiesis and organ weights in rats.
J Appl Toxicol 4:145-149.

Cho SH, Yun Dr (1988) Experimental study on the effects of acute carbon monoxide intoxication during pregnancy on fetal growth in rat.
Seoul J Med 29:55-63.

Chiodi H, Dill DB, Consolazio F et al (1941) Respiratory and circulatory responses to acute carbon monoxide poisoning.
Am J Physiol 134:683-693.

Corya BC, Black MJ, McHenry PL (1976) Echocardiographic findings after acute carbon monoxide poisoning.
Br Heart J 38: 712-7

Culver B, Norton S (1976) Juvenile hyperactivity after acute exposure to carbon monoxide.
Exp. Neurol 50:80

Dempsey SL, O'Donnell JJ, Hoft JT (1976) Carbon monoxide retinopathy.
Am J Ophthalmol 82:692-693.

Dwyer EM, Turino GM (1989) Carbon monoxide and cardiovascular disease.
New Engl J Med 321:1474-1475.

Ekblom B, Huot R (1972) Response to submaximal and maximal exercise at different levels of carboxyhemoglobin.
Acta Physiolog Scand 86:474-483.

Fedrick J, Alberman ED, Goldstein H (1971) Possible teratogenic effects of cigarette smoking.
Nature 231:529-530.

Fechter LD, Young JS, Carlisle L (1988) Potentiation of noise induced threshold shifts and hair cell loss by carbon monoxide.
Hearing Research 34:39-48.

Fechter LD, Annau Z (1977) toxicity of mild prenatal carbon monoxide exposures.
Science 197:680-682.

Finlay J, Van Beck A, Glover JL (1977) Myonecrosis complicating carbon monoxide poisoning.
J Trauma 17:536-540.

Findlay GH (1988) Carbon monoxide poisoning: Optics and histology of skin and blood.
British J of Dermatology 119: 45-51.

Goldbaum LR, Ramiraez RG, Abraham KB (1975) What is the mechanism of carbon monoxide toxicity?
Aviat Space Environ Med 46:1289-1291.

Goujard J, Rumeau C, Schwartz D (j1975) Smoking during pregnancy, stillbirth and abruptio placentae.
Biomedicine 23:20-22.

Ginsberg MD (j1985) Carbon Monoxide Intoxicaton: Clinical Features, Neuropathology and Mechanisms of Injury.
Clinical Toxicology 23: 281-288.

Haldane J (1895) The action of carbonic oxide on man.
J Physiol (Lond) 18:430.

Jain KK: Oxygen in Physiology and Medicine. Springfield, Thomas, 1989.

Hinderliter AL, Adams KF, Price CJ et al (1989) Effect of low-level carbon monoxide exposure on resting and exercise-induced ventricular arrythmias.
Arch Environ health 44:89-93.

Kelly JS, Sophocles GJ (1978) Retinal hemorrhages in subacute carbon monoxide poisoning: exposures in homes with blocked furnace flues.
JAMA 239:1515-1517.

King CE, Dodd SL, Cain SM (1987) Oxygen delivery to contracting muscle during hypoxic or CO-hypoxia.
J Appl Physiol 63:723-732.

Klausen K, Rasmussen B, Gjellerod L et al (1968) Circulation, metabolism, and ventilation during prolonged exposure to carbon monoxide and to high altitude.
Scand J Clin Lab Invest Suppl 103:26-38.

Koskela RS, Hernberg S, Karava R et al (1976) A mortality study of foundry workers.
Scand J Work environment health 2 (Suppl 1): 73-89.

Komatsu F (1970) Digest of science and labour. Cited in: Goldsmith JR: Carbon monoxide research - recent and remote.
Arch Environ Health 21: 118-120.

Komatsu E (1959) Shinshu myocarditis - a type of myocarditis caused by CO. In: Proceedings of the 15th General Assembly of the Japan Medical Congress, Tokyo, 1-5 April, 1959. pp 343-348 (in Japanese).

Kuller LH, Radford EP (1983) Epidemiological bases for the current ambient carbon monoxide standards.
Environmental Health Perspectives 52: 131-139.

Laas R, Inloffstein J, Meyerhof S (1983) Cerebral infarction due to carotid occlusion and carbon monoxide exposure.
J Neurol Neourosurg Psychiat 46:756-676.

Lahiri S, Mulligan E, Nishino T et al. (1981) Relatie responses of aortic body and caroitd boy chemoreceptors to carboxyhemoglobinemis.
J Appl Physiol (Resp. Environ Exercise Physiol 50: 580-6.

Lin H, McGrath JJ (1988) Vasodilating effect of carbon monoxide.
Drug and Chemical Toxicology 11:371-385.

Lewey FH, Drabkin DL (1944) Experimental chronic carbon monoxide poisoning of dogs.
Am J Med Sc 208:502-511.

Marklund Sl (1985) Oxygen toxicity and protective systems.
Clin toxicol 23:289-298.

Makishima K, Keane WM, Vernose GV et al (1977) Hearing loss of a central type secondary to carbon monoxide poisoning.
Trans Am Acad Ophthalmol Otolaryngol 84:452-57.

McGrath JJ (1984) The Effects of Carbon Monoxide on the Heart: An In Vitro Study.
Pharmacology Biochemistry & Behavior 21: 99-102.

Mylynshyn MJ, Cain CM, Villeneuve SM et al (1988) circulatory and metabolic responses to carbon monoxide hyposia during -adrenergic blockade.
Am J Physiol 255: H77-H84.

Meyer MB, Jonas BS, Tonascia JA (1976) Perinatal events associated with smoking.
Amer J Epidemiol 103:464-474.

Montgomery MR, Rubin RJ (1973) Oxygenation during inhibition of drug metabolism by carbon monoxide or hypoxic hyopxia.
J Appl Physiol 35:505-509.

Müller G, Pankow D (1978) Ceruplasmin activity in rat serum after carbon monoxide poisoning, methemoglobinemia and repeated bleeding.
Toxicol Lett 1:275-278.

Myers RAM, Snyder SK, Majeus TC (1985) Cutaneous blisters and carbon monoxide poisoning.
Ann Emerg Med 14:603-606.

Myers RAM, Linberg SE, Cowley RA (1979) Carbon Monoxide Poisoning: The Injury and Its Treatment.
JACEP 8:479-484.

Nag AC, Chen KC, Cheng M (1988) Effect of carbon monoxide on cardiac cells in culture. Am J Physiol 255:C292-C296.

Pankow D, Ponsold W, Glatzel W et al (1975) Combined effects of carbon monoxide plus sodium nitrite on carbon tetrachloride on the sciatic motor conduction velocity of rats. Int Arch Occupat Health 35:81-88.

Pappenheimer JR (1984) Hypoxic insomnia: effects of carbon monoxide and acclimatization. J Applied Physiol: Respiratory, Environmental + Ex. Physiol. 57: 1696-1703.

Pankow D, Ponsold W (1978) Effect of single and repeated carbon monoxide intoxications on urinary cathecholamine excretion in rats. Acta Biol Med Germ 37: 1589-1593.

Pankow D, Ponsold W (1981a) Zum Mechanismus der durch Kohlenmonoxideinwirkung induzierten Erhöhlung der Gesamthämoglobinkonzentration. Z Ges Hyg 27:208-213.

Pankow D, Ponsold W (1981b) Einfluss von Fremdstoffen auf die Kohlenmonoxid-induzierte Polyglobulie. Z Ges Hyg 27:351-357.

Pankow D, Haerting J, Ponsold W et al (1983) Seasonal variations of carbon monoxide induced polycythemia in rats. Biomed biochim Acta 42: 255-263.

Pankow D, Ponsold W, Müller E et al (1984) Adaptive Reaktionen im Knochenmark von Ratten nach akuter Kohlenmonoxidintoxikation. Kriminalistik und forensiche Wissenschaften 53:177-180.

Prinay F, DuJarin D, Deroanne R et al (1971) Muscular exercise during intoxication by carbon monoxide. J Appl Physiol 31:573-575.

Patterson CA, Smith E, Picket A (1938) Testes and hypophyses in gassed male rats. Proc Soc Exp Med Biol 38:455-466.

Penney DG, Barthel BG, Skoney JA (1984) Cardiac compliance and dimensions in carbon monoxide-induced cardiomegaly. Cardiovascular Res 18:270-276.

Schneiderman G, Goldstick TK (1978) Carbon Monoxide-Induced Arterial Wall Hypoxia and Atherosclerosis. Atherosclerosis 30: 1-15.

Shiotsuka RN, Drew RT, Wehner RW (1984) Carbon Monoxide Enhances Development of Hypertension in Dahl Rats. Toxicology and Applied Pharmacology 76: 225-233.

Smith JR, Landaw SA (1978) Smoker's Polycythemia. NEJM 298:6-10.

Seppänen A (1977) Physical work capacity in relation to carbon monoxide inhalation and tobacco smoking. Ann Clin Res 9:269-274.

Shapiro AB, Maturen A, Herman G et al (1989) Carbon monoxide myonecrosis.
Vet Hum Toxicol 31:136-137.

Sokal JA, Majka J, Palus J (1984) The content of carbon monoxide in the tissues of rats intoxicated with carbon monoxide in various conditions of acute exposure.
Arch Toxicol 56:106-108.

Sokal JA, Kralkoska E (1985) The relationship between exposure duration, carboxyhemoglobin, blood glucose, pyruvate and lactate and the severity of intoxication in 39 cases of acute carbon monoxide poisoning in man.
Arch Toxicol 57: 196-199.

Snyder RD (1970) Carbon monoxide intoxication with peripheral neuropathy.
Neurology 20: 177-180.

Spencer TD (1962) Effect of carbon monoxide on man and canaries.
Ann Occupational Hyg 5:231-240.

Stern FB, Halperin WE, Hornung RW et al (1988) Heart disease mortality among bridge and tunnel officers exposed to carbon monoxide.
Am J Epidemiol 128: 1276-1288.

Syvertsen GR, Harris JA (1973) Erythropoietin production in dogs exposed to high altitude and carbon monoxide.
Am J Physiol 225:293-299.

Tachi N, Aoyama M (1983) Effect of cigarette smoke and carbon monoxide inhalation by gravid rats on the conceptus weight.
Bull. Environ. Contam. Toxicol. 31 (1983) 85-92.

Thom SR (1988) Experimental carbon monoxide-mediated brain lipid peroxidation and the effect of oxygen therapy.
Ann Emergency Med 17:403.

Vanoli E, DeFerrari GM, Stramba-Badiale M et al (1989) Carbon monoxide and lethal arrythmias in conscious dogs with healed myocardial infarction.
Am Heart J 117:348-357.

Vogel JA, Gleser MA, Wheeler RC et al (1972) Carbon monoxide and physical work capacity.
Arch Environ Health 24:196-203.

Wells LL (1933) The placental effects of carbon monoxide on albino rats and the resulting neuropathology.
biologist 15:80-81.

Wenstrup J, Kjelsen K, Astrup P (1969) Acceleration of spontaneous intimal subintimal changes in the rabbit aorta by a prolonged moderate carbon monoxide exposure.
Acta Pathol Microbiol Scand 75:353-362.

Zorn H (1972) Der Sauerst off-Partialdruck im Hirngewebe und in der Leber bei subtoxischen Kohlenmonoxid-Konzentrationen.
Staub Reinhalt Luft 32:151-155.

# Chapter 6

# PATHOLOGY OF CARBON MONOXIDE POISONING

## INTRODUCTION

The pathophysiology and lesions resulting from CO poisoning have been partly described in Chapter 5. The lesions are listed in Table 6-1. The results of autopsy of 567 cases of CO poisoning from the files of the Armed Forces Institute of Pathology (Washington, DC) have been reviewed by Finck (1966). This remains a classical source of information on this subject. The brain is the organ most affected by CO poisoning.

TABLE 6-1
PATHOLOGICAL LESIONS OF VARIOUS ORGANS IN CO POISONING

| *Organ* | *Important lesions* |
|---|---|
| Brain | Acute: Cerebral edema and hemorrhages<br>Chronic: Necrotic lesions in basal ganglia and demyelination. |
| Heart | Acute: Myocardial necrosis<br>Chronic: Atherosclerosis coronary arteries, myocardial infarcts. |
| Lungs | Primary: Pulmonary edema<br>Secondary: Aspiration pneumonia in comatose patients |
| Liver | Lobar necrosis with chronic repeated exposure |
| Kidney | Parenchymatous degeneration leading to necrosis |
| Muscles | Intramuscular hemorrhages, swelling, and rhabdomyolysis |
| Bone | Marrow hypertrophy in chronic CO-hypoxia |
| Skin | Erythema, blisters, and gangrene |

## NEUROPATHOLOGY

### Gross appearance of the brain at autopsy

The brain appears to be pink in color and swollen. Petechial hemorrhages may be noted on the surface. Various herniations may be noted during removal of the brain from the calvarium.

In cases of chronic CO poisoning or those cases of acute poisoning where the interval between exposure and death is long, cerebral atrophy is noted.

### Pathophysiology of brain lesions

Brain lesions of CO poisoning are usually considered to be due to hypoxia.

Clinical instances of "pure hypoxia" are rare and many investigators consider CO intoxication to represent cerebral hypoxia aggravated by relative ischemia as the lesions are similar to those induced by other forms of hypoxia/ischemia. Of the cells of the CNS, the astrocytes are more sensitive than neurons to the CO effect. Small amounts of dissolved CO can inhibit cellular respiration although physiologically adequate amounts of oxygen may be present (Walum *et al.* 1985).

Putnam *et al.* (1931) showed that CO damages the blood brain barrier, particularly in the cerebral white matter, where the venous drainage pattern predisposes to focal edema. This may lead to hypoxia and set up the vicious circle: hypoxia $\rightarrow$ edema $\rightarrow$ hypoxia. Delayed neurological deterioration can occur following anoxia (Plum *et al.* 1962) due to other causes and can explain similar deterioration after CO poisoning, in absence of COHb elevation.

Siesjö (1985) believes that the mechanism of delayed neurological toxicity is based on several reactions triggered by increased calcium concentrations in the cell, which persist long enough to produce alterations in the cell function and delayed neurological damage.

Ginsberg (1985) studied the pathomechanism of CO lesions in anesthetized Rhesus monkeys exposed to 0.2% CO and leading to COHb levels of 75%. Bilateral lesions of the central white matter of the brain, found in these animals, were remarkably similar to the lesions found in human CO poisoning. There was necrosis of the white matter but sparing of the cortical U-fibers and of the axons. Analysis of his physiological data revealed that marked carboxyhemoglobinemia, although a necessary precondition, was itself insufficient to produce white matter lesions. Lesion size correlated positively with the degree of hypotension and systemic acidosis. Okeda *et al.* (1981) studied the effects of various physiological factors contributing to the pathogenesis of CO-induced lesions of the white matter. He concluded that the selective damage is due to the following:

1. Coexistence of the initial phase of increase in CBF and a subsequent decrease of CBF.

2. Anatomical findings that the cerebral white matter is supplied by its own long nutrient arteries with smaller amounts of capillary beds and thinner media the subarachnoid arteries.

Brucher (1967) in his detailed studies on CO poisoning and anoxia states that:

1. CO-induced encephalopathy is not fundamentally different from other anoxic encephalopathies.

2. As in other anoxic lesions, the following factors may take part in the genesis of CO-encephalopathy: hypoxidosis, edema, circulatory disturbances, and histotoxic action.

3. Each of these factors can become active at successive times and to a variable degree. This explains the polymorphism of the lesions and the clinical features.

4. The severity of anoxic encephalopathy is not always proportional to the severity of the anoxia.

There are three stages in the evolution of the brain lesions:

1. In immediate death after CO poisoning, there are petechial hemorrhages throughout the brain but no cerebral edema.

2. In patients who die within hours or days after poisoning, cerebral edema is present. There is necrosis of globus pallidus and substantia nigra.

3. In patients who die days or weeks later from delayed sequelae of CO poisoning, edema has usually disappeared. Degenerative and demyelinative changes are usually seen.

**Location of lesions**

1. **Basal ganglia.** Lesions in this location were found in 16 of the 22 fatal studied by Lapresle and Fardeau (1967). The lesions are usually located within the globus pallidus and are bilateral (Fig. 6-1). Histologically, the necrotic areas show diapedetic hemorrhages, hemosiderin-laden macrophages, and calcium deposits in the walls of the blood vessels in the necrotic zone. The alterations in the basal ganglia are not recognized unless the patient survives the acute CO poisoning by more than 24 hours.

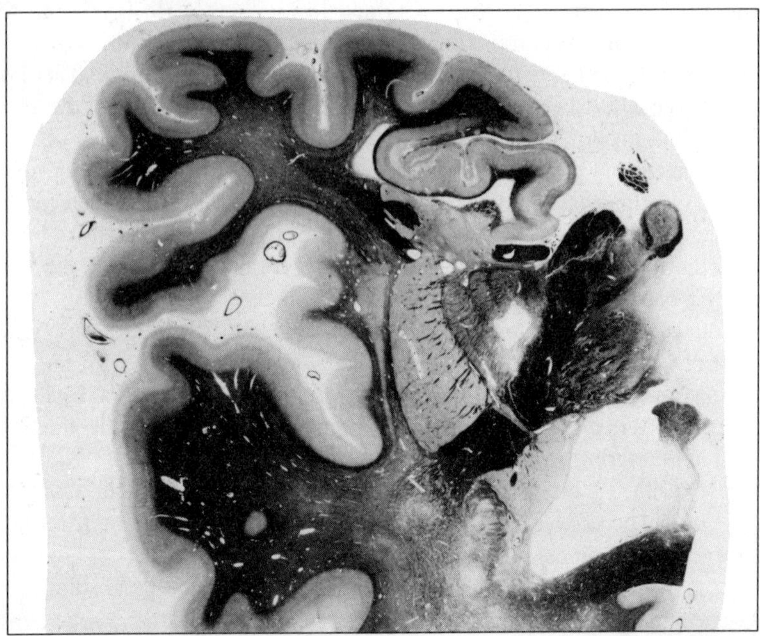

Fig. 6-1: Anoxic CO leukoencephalopathy with typical spongy appearance of the white matter in the frontal lob (left) and occipital lobe (right). Normal gray matter. (Paraffin, Kluver-Barrera). Reproduced from: Brucher JM: Leukoencephalopathies in anoxic-ischemic processes. In, Koetsier JC (ed): Handbook of Clinical Neurology, vol 3 (47), Amsterdam, Elsevier Science Publishers, pp.527-549. (By permission of the author and the publishers).

2. **Ammon's horn (hippocampus).** The lesions in this location are found in about half of the cases of CO poisoning. There are necrotic with hemorrhages and calcium deposits. Either the entire horn or only the Sommer's sector is involved.

3. **White matter.** Alterations in white matter are a conspicuous and common sequel of CO poisoning. They take one or more of the following three forms: (i) Discrete perivascular foci of myelin destruction throughout the centra semiovale, the corpus collosum, the internal capsule, and the optic tracts. Within each lesion the breakdown of myelin and axis cylinders is followed by proliferation of lipid phagocytes and of fibrous astrocytes.

(ii) Diffuse anoxic leukoencephalopathy. The destruction of myelin fibers is diffuse and widespread. It may extend from the frontal to the temporal and occipital poles, extending along the axes of the convolutions but sparing the subarcuate fibers. The demyelinated areas may present a "moth eaten" or spongy appearance (Fig. 6-2).

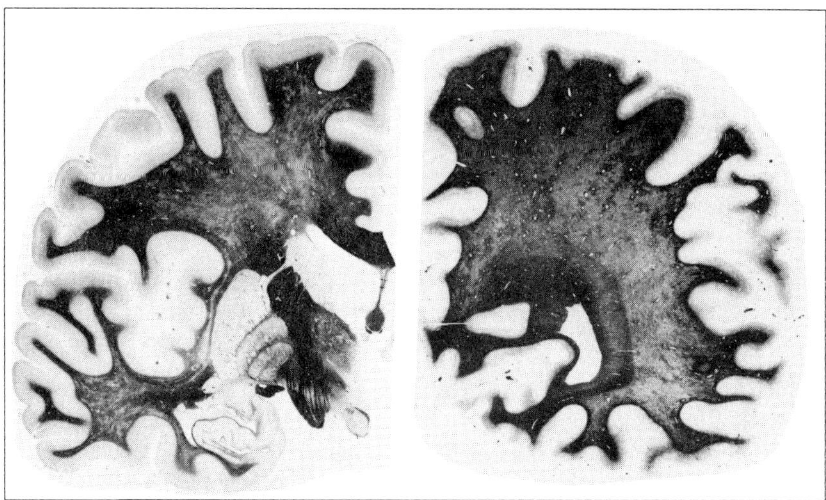

Fig. 6-2: A case of carbon monoxide poisoning with coma, recovery and delayed recurrence of coma. Cystic necrosis the upper part of the pallidum on both sides, and diffuse, patchy demyelination in centrum semiovale and temporal white matter. (Paraffin, Spielmeyer). Acknowledgement the same source as in fig. 6-1.

(iii) Frank demyelination. This may take the form of plaques. An extensive patch of demyelination in the frontal lobe of a patient who died of CO poisoning is shown in Figure 6-3 Histologically there is relative sparing of axis cylinders in such lesions.

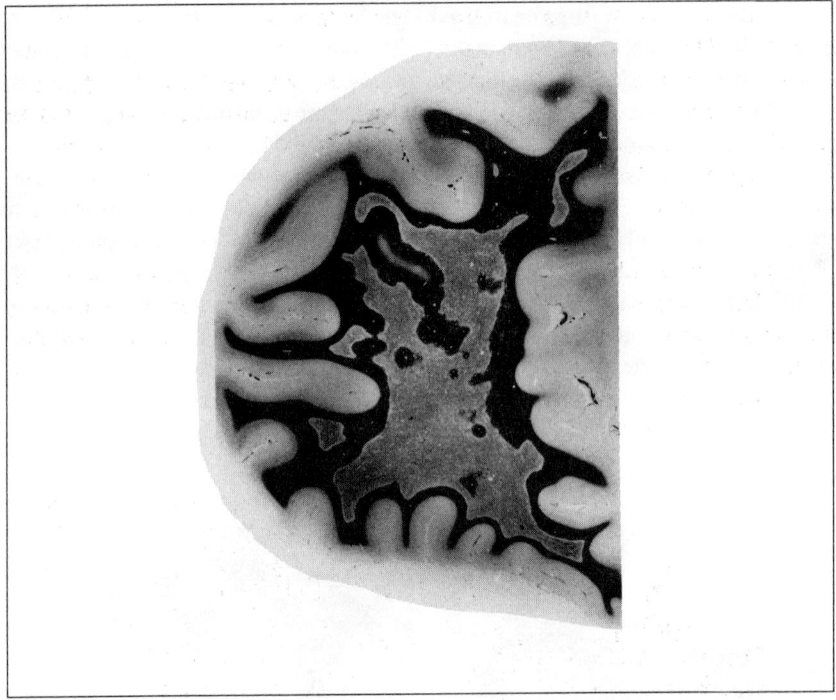

Fig. 6-3: Section of frontal lobe from a case of carbon monoxide poisoning showing extensive demyelination. From, Hart IK, Kennedy PGE, Adams JH *et al:* Neurological manifestations of carbon monoxide poisoning. Postgraduate Med J 1988;64:213-215. By permission.

Such lesions were first described by Grinker (1925) who correlated them with a biphasic clinical picture. Jacob (1939) was the first to suggest that degeneration in the white matter rather than in the gray matter was the neuropathological substrate in those patients where the neuropsychiatric symptoms followed the initial intoxication after a period of apparent recovery. Other examples of this relationship were the two cases described by Hsü and Ch'eng (1938). Courville (1957) also described the residual lesions in the cerebral white matter as a delayed sequel of exposure to carbon monoxide.

Diffuse anoxic encephalopathy is sometimes associated with clearly defined areas of total necrosis (true softening) disseminated throughout the centrum semiovale and associated with perivascular hemorrhagic suffusions (Hiller 1924).

4. **Diffuse gray matter lesions.** These lesions may be found in the cerebral and cerebellar cortices and various gray matter nuclei. In contrast to me view of Jacob (1939), Dooling and Richardson (1976) consider gray matter lesions to be the basis of relapsing clinical course whereas white matter lesions are more characteristic of an uninterrupted clinical course.

## PERIPHERAL NEUROPATHY

Peripheral neuropathy as a sequel of CO poisoning is rare. Grunett and Petajan (1976) loss of normal axonal and Schwann cell structure by electron microscopy in rats poisoned with CO.

## THE HEART

Lesions of the heart in experimental animals poisoned with CO have been reviewed by Pankow (1981). The lesions vary from hyperemia to frank necrosis and hemorrhages. Similar lesions are seen in human patients who die of CO poisoning.

## LUNGS

Finck (1966) reported pulmonary edema in 64% and pulmonary hemorrhages in 10% of cases at autopsy. Aspiration pneumonia may be found in patients who have been comatose prior to death.

## LIVER

Experimental evidence for the liver lesions after CO poisoning has been reviewed exhaustively by Pankow (1981). Necrotic lesions have been demonstrated in e liver parenchyma after prolonged CO exposures.

## KIDNEYS

Various lesions in the kidneys seen in fatal CO poisoning are:

1. Hyperemia
2. Edema of the interstitial tissues.
3. Parenchymatous degeneration
4. Fatty deposits in the epithelial cells of the ascending tubes of Henle.

## MUSCLES

The following lesions have been noted:

1. Intramuscular hemorrhages
2. Interstitial edema
3. Swelling
4. Loss of striations to full degeneration of the muscle fibers.
5. Rhabdomyolysis.

## SKIN

The following skin lesions have been noted:

1. Skin blisters and blebs
2. Edema
3. Damage to the eccrine sweat acini
4. Erythma
5. Necrosis and gangrene
6. Loss of hair

Myers *et al.* (1985) postulate that the skin lesions occur as a result of high COHb levels and are not due to any local effect or burns. These lesions demonstrate intraepidermal and subepidermal vesicles with occlusion of intraepidermal sweat pores; these vesicles ultimately coalesce to form subepidermal bullae (Leavell *et al.* 1969).

The classical cherry red color of the skin in fatal cases of CO poisoning may be obvious. The optical features of the skin at autopsy have been described by Findlay (1988).

## FETAL DAMAGE IN PREGNANCY

The effect of CO on the fetus has been described in Chapter 5. In severe cases of CO poisoning in pregnancy, the maternal mortality is about 33% but the fetal mortality is over 66%. Ginsberg and Myers (1974) have shown fetal brain damage in rhesus monkeys exposed to CO leading to 60% COHb concentration in only those fetuses where the arterial oxygen content fell below 2.0 ml/100 ml for at least 45 min during maternal CO intoxication. Malformations of the limbs have also been reported following fetal intoxication during the first trimester.

Cause of death by CO poisoning is often investigated by coroners and forensic medical specialists. The circumstances surrounding the exposure to CO are important. This information is important in cases of accidental CO poisoning and may prevent similar further happenings at the site of accident.

Procedures for determining COHb in the post-mortem blood are described in Chapter 8. In delayed deaths, the proof of exposure to CO depends on the analysis of the air sample taken at the scene. The lesions encountered do not allow a specific diagnosis of CO poisoning. Other causes of death and contributing causes should be looked for. Blood alcohol determination is important. Cyanide poisoning is an important factor in smoke inhalation injuries.

The histopathological change in the brain from other causes such as cardio-respiratory arrest and hyperglycemia.

## REFERENCES

Brucher JM: Neuropathological problems posed by carbon monoxide poisoning and anoxia. In, Bour H and Ledingham IM (eds) Carbon Monoxide poisoning. Amsterdam, Elsevier Publishing company, 1967, pp 75-100.

Caravati EM, Adams CJ, Joyce SM et al. (1988) Fetal Toxicity Associated With Carbon Monoxide Poisoning. Annals of Emergency Medicine 17: 714-717.

Courville CB: The process of demyelination in the central nervous system. IV. Demyelination as a residual of carbon monoxide asphyxia. J Nerv Ment Dis 1957; 125:534-546.

Doolin EC, Richardson EP Jr: Delayed encephalopathy after strangling. Arch Neurol 1976; 33:196.

Finck PA: Exposure to carbon monoxide: Review of literature and 567 autopsies. Military Med 1966; 131: 1513-1539.

Findlay GH: Carbon monoxide poisoning: Optics and histology of skin and blood. Brit J Dermatol 1988:119:45-51.

Ginsberg MD, Myers RE: Fetal brain damage following maternal carbon monoxide poisoning: An experimental study. Acta Obstet Gynec Scand 1974; 53:309-317.

Ginsberg MD; Carbon monoxide intoxication: Clinical features, neuropathology and mechanism of injury. Clin Toxicol 1985;23:281-288.

Grinker RR: Über einen Fall von Leuchtgasvergiftung mit doppelseitiger Pallidumerweichung und schwerer Degeneration des tiefen Grosshirnmarklagers. Zeitschrift Neurologie und Psychiatrie 1925;98:433-456.

Grunnet JH, Petajan JH: Carbon monoxide induced neuropathy in the rat. Arch Neurol 1976; 33:152-157.

Hiller F: Über die krankhaften Veränderungen im Zentralnervensystem nach Kohlenoxydvergi- tung. Zeitschrift Neurologie und Psychaitrie 1924;93:594-646.

Hsü YK, Ch'ng YL: Cerebral cortical myelinopathy in carbon monoxide poisoning. Brain 1938;61:384-392.

Jacob H: Über die diffuse Hemisphärenmarkerkrankung nach Kohlenoxydvergiftung bei Fällen mit klinisch intervallärer Verlaufsform. Zeitschrift Neurologie und Psychiatrie 1939;139:161-179.

Lapresle J, Fardeau M: The central nervous system and carbon monoxide poisoning. In, Bour H, Ledingham IM (eds): Carbon Monoxide Poisoning. Progress in Brain Research, vol 24. Amsterdam, Elsevier Publishing company 1967. pp 31-74.

Leavell UW, Farley CH, McIntyre JS: Cutaneous changes in a patient with carbon monoxide poisoning.
Arch Dermatol 1969;99:429-433.

Myers RAM, Snyder SK, Majerus TC et al: Cutaneous blisters and carbon monoxide poisoning. Annals of Emerg Med 1985;14:603-606.

Newby MB, Roberts RJ, Bhatnager RK (1978) Carbon monoxide and hypoxia-induced on catecholamines in the mature and developing rat brain.
J Pharmacol Exp. Ther 206: 61.

Ogawa M, Katsurada K, Sugimoto T et al (1974) Pulmonary edema in acute carbon monoxide poisoning.
Int Arch Arbeitsmed (Berlin) 33:131-138.

Okeda R, Funata N, Takano T et al (1981) The Pathogenesis of Carbon Monoxide Encephalopathy in the Acute Phase - Physiological and Morphological Correlation.
Acta Neuropathol 54: 1-10.

Pankow D: Toxikolgie des Kohlenmonoxids.
Berlin, VEB Verlag, 1981. pp 95-103.

Plum F, Posner JB, Hain RF: Delayed neurological deterioration after anoxia.
Arch Int Med 1962;110:18-25.

Putnam TJ, McKenna JB, Morrison LR: Studies in multiple sclerosis.I. Histogenesis of experiemental sclerotic plaques and their relation to multiple sclerosis.
JAMA 1931;97:1591-1596.

Sawada Y, Takahashi M, Ohash N et al: Computerized tomography as an indication of long term outcome after acute carbon monoxide poisoning.
Lancet 1980; i:783-784.

Schwartz A, Hennerici M, Wegner OH: Delayed choreoathetosis following acute carbon monoxide poisoning.
Neurology 1985;35:98-99.

Siesjö BK (1985) Carbon monoxide poisoning: mechanism of damage, late sequelaate and therapy.
Clin Toxicol 23:247-248.

Walum E, Varnbo I, Peterson A (1985) Effects of Dissolved Carbon Monoxide on the Respiratory Activity of Perfused Neuronal and Muscle Cell Cultures.
Clincial Toxicology 23:299-308.

# Chapter 7

## CLINICAL FEATURES OF CARBON MONOXIDE POISONING

### SIGNS AND SYMPTOMS OF CARBON MONOXIDE POISONING

The signs and symptoms of CO poisoning are non-specific and involve most of the body systems. Classically, the signs and symptoms are described according to COHb levels as shown in Table 7-1. In practice, the COHb levels do not necessarily correlate with the severity of the clinical manifestations.

TABLE 7-1
DEGREE OF SEVERITY OF CO POISONING, COHb LEVELS AND CLINICAL FEATURES

| *Severity* | *COHB level* | *Clinical Features* |
|---|---|---|
| Occult | >5% | 1. No apparent symptoms<br>2. Psychological deficits on testing |
| | 5%-10% | 1. Decreased exercise tolerance in patients with chronic obstructive pulmonary disease.<br>2. Decreased threshold for angina and claudication in patients with atherosclerosis.<br>3. Increased threshold for visual stimuli |
| Mild | 10%-20% | 1. Dyspnea on vigorous exertion.<br>2. Headaches, dizziness.<br>3. Impairment of higher cerebral function<br>4. Decreased visual acuity |
| Moderate | 20%-30% | Severe headache, irritability, impaired judgement visual disturbances, nausea, dizziness, increased respiratory rate. |
| | 30%-40% | Cardiac disturbances, muscle weakness, vomiting, reduced awareness. |
| Severe | 40%-50% | Fainting on extertion, mental confusion |
| | 50%-60% | Collapse convulsions, paralysis. |
| Very severe | 60%-70% | Coma, frequently fatal in a few minutes. |
| | over 70% | Immediately fatal. Respiratory and cardiac arrest. |

## Classification of Severity of CO Poisoning on Clinical Basis

The severity of CO poisoning depends on the dose of CO and the duration of exposure. A practical classification of the severity of CO poisoning is shown in Table 7-2. The degree of severity of CO poisoning corresponds more closely to the blood lactate levels than to COHb levels (Sokal and Kralkowska 1985).

TABLE 7-2
CLASSIFICATION OF SEVERITY OF CO POISONING
ON BASIS OF CLINICAL FEATURES

| *Clinical degree of intoxication* | *Specification of symptoms* |
|---|---|
| I<br>mild | Headache, vomiting, tachycardia; no disturbances of consciousness |
| II<br>moderate | Disturbances or loss of consciousness without other neurologicial symptoms, tachycardia; nociceptive reflexes still intact |
| III<br>severe | Loss of consciousness, intense muscular tonus, pathological neurological symptoms, tachycardia and tachypnoea; circulatory and respiratory disurbances not observed |
| IV<br>Very severe | Loss of consciousness, clinical signs of central nervous system damage, circulatory and respiratory disturbances |

## Chronic CO Poisoning

Symptoms of chronic exposure to CO are vague. Headache is the most constant, often accompanied by nausea and vomiting in severe cases. Muscular weakness and tiredness on moderate exertion and other symptoms suggestive of myocardial insufficiency may be elicited on questioning the patient. A list of symptoms with their frequency is shown in Table 7-3.

TABLE 7-3
SUBJECTIVE SYMPTOMS OF CHRONIC CO POISONING

| *Symptom* | *Frequency (%)* |
|---|---|
| Fatigue | 92 |
| Headache | 87 |
| Dizziness | 69 |
| Sleep disturbances | 66 |
| Cardiac symptoms | 62 |
| Apathy | 54 |
| Nausea, vomiting | 42 |
| Memory disturbances | 40 |
| Reduced libido | 22 |
| Loss of appetite | 17 |

The exact definition of chronic CO poisoning is difficult as the levels of environmental CO, the level of COHb in blood, and the duration of exposure required to produce subjective and objective symptoms are variable. The term "residual CO poisoning" is preferred by some authors (Krüger *et al.* 1960).

## CLINICAL DIAGNOSIS

History of exposure to CO and the circumstances in which the patient is found are more helpful in the diagnosis than clinical examination. Marked tachycardia and tachpnea are common. The classical "cherry red" color of the skin is rare and seen usually in fatal cases. The patients are more likely to appear pale or cyanotic (Meigs and Hughs 1952). Retinal hemorrhages when present may suggest the diagnosis. Kelly and Sophocleus (1978) reported 12 patients with subacute poisoning who presented with headache and "flu" like symptoms. All of them had retinal changes. Ferguson *et al.* (1985) reported three family members with acute CO poisoning resulting from a single accidental exposure. The father had a normal fundus but the mother had scattered superficial retinal hemorrhages. The child had the most pronounced findings; large subinternal limiting membrane hemorrhages, peripapillary hemorrhages, severe venous tortuosity, and disc edema. The fundus parallel the variable clinical findings of CO poisoning.

## PITFALLS IN THE CLINICAL DIAGNOSIS OF CO POISONING

The following pitfalls should be noted:

1. Clinical signs and symptoms of CO poisoning do not always correspond to COHb levels.
2. Cherry red color of the skin and the lips, usually considered to be a classical sign, is not present when the COHb levels are below 40% and there is cyanosis due to respiratory depression. In practice this sign is seen only rarely.
3. Some of the symptoms are aggravated by pre-existing disease such as intermittent claudication.
4. Tachpnea is frequently absent because the carotid body is presumably responsive to the oxygen partial pressures rather than the oxygen content.
5. Examples of frequent misdiagnoses of CO poisoning are (Barret 1983):
   - i. Psychiatric illness
   - ii. Migraine headaches
   - iii. Stroke
   - iv. Acute alcohol intoxication or delerium tremens
   - v. Heart disease
   - vi. Food poisoning

Littlejohns (1989) has described the case of a young lady who tried to commit suicide in her car by connecting the exhaust to the inside by a hose. The diagnosis of CO poisoning was missed and the patient was treated by electroconvulsive therapy. Persistent cognitive deficit led to the diagnosis of CO poisoning.

## OCCULT CO POISONING

Occult CO poisoning is a syndrome of headache, fatigue, dizziness, paresthesias, chest pains, palpitation and visual disturbances associated with chronic CO exposure (Kirkpatrick 1987). Headache and dizziness are early symptoms of CO poisoning and occur at COHb concentrations of 10% or more. Heckerling *et al.* (1988) have shown that, among patients taken to an emergency department during the winter heating season with complaints of headache or dizziness, 3%-5% had COHb levels in excess of 10%. They were unaware of exposure to toxic levels of CO in their home prior to the visit to the emergency department. A retrospective decision to determine COHb levels only in those patients who used gas stoves for space heating, correctly identified most patients with occult CO poisoning.

### Less Known Manifestations of CO Poisoning

1. Polycythemia. This can occur in chronic CO poisoning but when a patient presents with polycythemia in the absence of a history of CO exposure, this cause is usually overlooked. DiMarco (1988) described the case of a 62 year old woman who presented with headache, dizziness and difficulty in concentration. She had polycythemia and mild hypertension. She received various treatments for 1 1/2 years, but the diagnosis of CO poisoning was missed. Finally her Hb was shown to be 37% and a leak in the gas pipe at home was identified.

2. Thrombotic thrombocytopenic purpura (TTP). Stonesifer *et al.* (1980) described a patient with CO poisoning where the diagnosis was missed initially. The patient was arrested for unruly behavior and became comatose in jail. On admission to hospital, his COHb was 20%. He developed respiratory distress and died. Post-mortem findings were consistent with the diagnosis of TTP and CO-encephalopathy. Vascular damage induced by CO with platelet accumulation may initiate TTP and patients with CO poisoning should be evaluated for TTP.

3. Acute pancreatitis (Taha *et al.* 1989). Serum amylase levels have been shown to be elevated in patients with CO poisoning. The values returned to normal after HBO treatments.

## DIAGNOSIS OF CO POISONING IN CHILDREN

Carbon monoxide presents special risks to infants. The non-specific presenting symptoms make it difficult to establish a diagnosis of CO poisoning which is frequently missed. The special features of fetal hemoglobin may lead to greater tissue hypoxia than in older individuals. Finally an altered state of consciousness, with or without convulsions, is a common presentation of many diseases of children. The environmental causes may not be considered in the absence of a history of exposure and when no adult members of the household show any symptoms of CO poisoning.

When unexplained neurological symptoms occur in an infant who has been a passenger in a car, COHb determinations should be done and CO poisoning should be considered in the differential diagnosis (Venning *et al.* 1982). Gamelli and Cattani (1985) described three children who presented with signs and symptoms of gastroenteritis and subsequently diagnosed to have CO poisoning which was traced to a faulty gas pipe at home.

Children are more susceptible to CO poisoning and show severe manifestations at COHb levels that may be asymptomatic in the adults. Chronic CO exposure may lead to mental retardation and behavior problems in children (Binder and Roberts 1980).

## NEUROPSYCHOLOGICAL EFFECTS OF RAISED COHb LEVELS

There are several studies of the effect of raised COHb on psychomotor and cognitive function. These are summarized in Table 7-4. Because of inadequate methodology, we disregard the results of an older study (Dorcus and Weigland 1929) which failed to find any deterioration of cognitive or psychomotor function with COHb levels as high as 25%-35%. Tests of tracking performance have been reported to show no impairment at 20% COHb (Schaad *et al.* 1983).

Bunnet and Hovarth (1988) tested cognitive function in relation to physical activity under CO exposure to simulate conditions likely to be found at work places. They found that elevation of COHb to 10% had no effect on spatial processing, short-term memory, simple reaction time, and psychomotor tacking. In contrast, performance was impaired when COHb level was increased following physical work ($VO_2$max 60%).

In order to simulate driving conditions under CO exposure, Ray *et al.* (1970) conducted driving tests in healthy volunteers with varying degrees of exposure to CO. They found that at 10% COHb level, the time required to respond to velocity detection tasks was significantly prolonged. This study is significant in relation to traffic safety because 10% COHb level can be reached easily in a smoker driving in congested traffic.

Visual perception has also been studied after exposure to CO. McFarland *et al.* (1944) reported loss of visual sensitivity associated with smoking. Vollmer *et al.* (1946) showed that flicker fusion frequency and visual perimetry were not affected by COHb levels as high as 22%. This study was simulated at 10,000 and 15,000 ft altitude and the hypoxic effects of these altitudes may have induced compensatory phenomena to mask the effects of CO. Halperin *et al.* (1959)

TABLE 7-4
NEUROPSYCHOLOGICAL SEQUELAE OF RAISED COHb LEVELS

| *Author and year* | *Subjects* | *COHb level or CO/ppm* | *Effects* |
|---|---|---|---|
| Lilienthal and Fugitt (1946) | Humans | 5%-10% COHb | Impairment in the Flicker Fusion Frequency test. |
| Trouton and Eysenck (1961) | Humans | 5%-10% COHb | Impairment of the precision of control.<br>Multiple limb incoordinations. |
| Schulte (1963) | Humans | 2%-5% COHb | Decrease in cognition and psychomotor ability. Increase in the number of errors and the completion time in arithmentic tests, t-crossing test and visual discrimination tests. |
| Beard and Wertheim (1967) | Humans | 90 min at 50 ppm or shorter time at 250 ppm (=COHb of 4%-5%) | Impaired ability to discriminate short time intervals. |
| | Rats | 100 ppm for 11 min | Disruption of ability to judge time (assessed by differential reinforcement at a low rate of response.) |
| Mikulka (1973) | Humans | 125-250 ppm briefly (COHb 6.6%) | No effect on time estimation<br>No disruption of tracking. |
| Hovarth et al (1973) | Humans | 111 ppm (COHb 6.6%) | Impairment of visual function. |
| Gliner et al (1983) | Humans | 100 ppm for 2.5 hr<br>Controls (room air) | Decreased arousal and interest with fatigue resulting in decrease in performance. |
| Schrott et al (1983) | Rats | 500 ppm-90 min (40% COHb)<br>850 ppm-90 min (50% COHb)<br>1200 ppm - 90 min (60% COHb) | Disruption of the rate at which the rats acquired a chain of response. |
| Yastrebov et al (1987) | Humans | 900 ∓ 20mg/m³ for 10 min (COHb of 10 ∓ 0.5%) | Impairment in a 2-dimensional compensatory tracking task combined with mental arithmetic. Symptoms of mild CO intoxication at COHb levels of 10% |

showed that visual function was impaired when COHb rose to 4-5%. Wiley (1989) has shown that there are no differences in visual function between smokers and non-smokers.

Studies of Beard and Wertheim (1967) show an alteration in behavioral performance with COHb levels below 5%. Efforts to reproduce these effects have not been uniformly successful. This study may be more relevant to impairment of vigilance which is the ability to detect small changes in the environment, changes which take place at tan unpredictable moment and require constant attention. Impairment of auditory vigilance has been shown at COHb levels of 3% (Groll-Knapp *et al.* 1972).

## NEUROLOGICAL EFFECTS OF ACUTE CO POISONING

Higher levels of COHb during acute exposure may lead to impairment of consciousness, coma, and convulsions. Bour *et al.* (1967) have reviewed the neurological features of 20 fatal cases of CO poisoning. They were all comatose with pyramidal signs, and often decerebratre hypertonia. There are no definite criteria of an unfavorable prognosis. Persistence of coma beyond first 48 hr is an indication of poor prognosis for recovery. Bokonjic (1963) has shown that the maximum period of CO-induced posthypoxic unconsciousness compatible with complete neurological recovery is 21 hr in patients under 48 yr of age. Complete recovery of mental function was not observed when CO-induced coma exceeded 15 hr in the older group or 64 hr in the younger group. Abbott (1972) reported the case of an 18 year old patient who remained comatose for 6 days following CO poisoning but made a complete recovery in 5 months. This patient was treated with both hyperbaric as well as normobaric oxygen.

## LATE NEUROLOGICAL SEQUELAE OF CO POISONING.

Most of the neurological and psychiatric manifestations of CO poisoning are late sequelae. These are listed in Table 7-5. The late complications of CO poisoning are also referred to as "secondary syndromes." The complications may develop a few days to three weeks after exposure to CO (Werner *et al.* 1985) and as late as two years, after apparently complete recovery from acute effects of CO poisoning. Neuropsychiatric symptoms are prominent in the late sequelae (Takahashi *et al.* 1987). These symptoms are listed in Table 7-6.

TABLE 7-5
DELAYED NEUROPSYCHOLOGICAL SEQUELAE OF CARBON MONOXIDE POISONING

*Psychological sequelae*

Memory disturbances
Diminised IQ
Temporospatial disorientation
Psychoses

*Neurological*

Ataxia
Dysgraphia
Movement disorders. Parkinsonism, choreoathetosis
Urinary incontinence
Apraxia. Ideomotor as well as constructional.
Convulsive disorders
Cortical blindness
Deafness (neural).
Peripheral neuropathy
EEG changes
Akinetic mutism
Gilles de la Tourette syndrome
Symptoms resembling those of multiple sclerosis
Dementia

TABLE 7-6
SYMPTOMS OF LATE NEUROPSYCHOLOGICAL SEQUELAE OF CARBON MONOXIDE POISONING *(After Min et al 1986).*

| SYMPTOM | NUMBER (%) OF PATIENTS |
|---|---|
| Psychiatric | |
| Apathy | 86 (100) |
| Disorientation | 86 (100) |
| Amnesia | 86 (100) |
| Hypokinesia | 82 (95) |
| Mutism | 82 (95) |
| Irritability, distractibility | 78 (91) |
| Apraxia | 65 (76) |
| Bizarre behaviors—silly smiles or frowning | 60 (70) |
| Manneristic behavior | 35 (41) |
| Irrational confabulatory talking | 26 (30) |
| Insomnia | 16 (19) |
| Depressed mood | 13 (15) |
| Delusions | 10 (12) |
| Echolalia | 2 (2) |
| Elated mood | 2 (2) |
| Neurologic | |
| Urinary and/or fecal incontinence | 80 (93) |
| Gait disturbance | 78 (91) |
| Glabella sign | 78 (91) |
| Grasp reflex | 75 (87) |
| Increased muscle tone | 74 (86) |
| Retropulsion | 62 (72) |
| Increased DTR | 19 (22) |
| Flaccid paralysis | 16 (19) |
| Tremor | 12 (14) |
| Dysarthria | 8 (9) |

The incidence of secondary syndromes varies from 10% to 30%. During a 3-year follow-up of patients exposed to CO, up to two-fifths developed memory impairment and one-third suffered deterioration of personality (Smith and Brandon 1973). In a collective review, Myers *et al.* (1985) found that 12% of patients poisoned by CO and treated by oxygen, still developed late sequelae 1-21 days (mean 5.7 d) after initial exposure. No such sequelae developed in patients treated by hyperbaric oxygen therapy.

Choi *et al.* (1983) reported that of the 2,360 victims of CO poisoning, delayed neurological sequelae were diagnosed in 11.8% of those admitted to the hospital and 2.75% of the total group. The lucid interval before appearance of neurological symptoms was 2-40 days (mean 22.4 d). The most frequent symptoms were mental deterioration, urinary incontinence, gait disturbance and mutism. The most frequent signs were masked face, glabellar sign, grasp reflex, increased muscle tone, short-step gait, and retropulsion. Most of theses signs indicate parkinsonism. There were no important contributing factors except anoxia and age. Previous associated disease did not hasten the development of sequela. Of the 36 patients followed for two years, 75% recovered within one year. Yun *et al.* (1987) reported an incidence of 41% of delayed neurological manifestations of CO poisoning. The incidence was higher in the older age groups and in females. It was also higher and proportional to the duration of exposure and of coma.

Ali-Cherif *et al.* (1984) described behavioral and mental activity disorders in two patients after CO poisoning. These patients had no motor deficits or movement disorders but had mental inertia and showed mental impairment on cognitive and memory testing. Pulst *et al.* (1984) described a case of Gilles de la tourette syndrome following CO poisoning. A previously healthy 58-year old man, following recovery from coma, developed tics in the head, fits of shouting and abnormal vocal utterances, in addition to signs of diffuse encephalopathy. CT scan showed a low density area in the basal ganglia and ventricular enlargement. In 290 cases of severe CO poisoning in Paris (Bour *et al.* 1966), 33.7% developed decerebrate rigidity. The usual duration of coma was 16 hours. However, one patient remained in coma for 5 days. Six of their patients had residual psychiatric syndromes.

Neuropsychological sequelae can occur after a symptom-free interval following acute CO poisoning. Funatsu *et al.* (1985) described the case of a 38 year old previously healthy man who developed impairment of drive and movement, disturbed memory, disorientation and abnormal gait two weeks following recovery from acute CO poisoning. He made a complete recovery after HBO treatments.

Quillam (1984) reported the case of a 63 year old man who presented with impaired consciousness after CO exposure and COHb level of 18%. He recovered completely after treatments with 100% normobaric oxygen but presented four weeks later with late neuropsychiatric sequelae. He deteriorated and died and post mortem examination showed changes of post-anoxic encephalopathy consistent with diagnosis of CO poisoning. Dordain *et al.* (1988) described a case of CO poisoning with coma who recovered completely and was discharged from hospital. He was readmitted 2 weeks later in an apathetic state and died. At autopsy, the brain showed typical changes of CO encephalopathy. Similar examples of remission and relapse have been reported previously (Garland and Pierce 1967, Nick *et al.* 1965, Garrel *et al.* 1970).

Larkin *et al.* (1976) recommend that all patients recovering from severe hypoxic insult should be followed up by periodical neuropsychological testing to detect and evaluate any cognitive or personality deficits.

## CLINICOPATHOLOGICAL CORRELATIONS OF CO POISONING

TABLE 7-7
CLINICO-PATHOLOGICAL CORRELATIONS IN CO POISONING

| *Organ involved* | *Manifestations* | *Pathology* | *Diagnostic procedures* |
|---|---|---|---|
| Brain | Acute: coma, seizures<br>Late: psychoses, dementia movement disorders | Cerebral edema, encephalopathy<br>Cerebral atrophy, basal ganglia lesions | EEG, CT scan, MRI<br>Neuropsychological testing, CT Scan |
| Heart | Angina, arrythmias | Myocardial ischemia, infarcts | ECG |
| Lungs | Dyspnea | Pulmonary edema | Chest X-rays, pulmonary function tests |
| Kidneys | Acute renal failure | Parenchymatous necrosis myoglobinuria | Urine examination, renal function tests |
| Skeletal muscles | Pain and weakness | Rhabdomyolysis | Serum creatine phosphatase |
| Eyes | Visual impairment | Retinal hemorrhages | Fundoscopy |
| Hearing | Hearing loss (neural) | Hypoxia of cochlear nerve | Audiometry |

Important clincopathological correlations of lesions in CO poisoning and the relevant laboratory investigations are shown in Table 7-7. Clinicopathological correlations are easy to make if cerebral lesions are visualized by imaging techniques. Koyabashi *et al.* (1984) have compared the CT findings of interval form of carbon monoxide with neuropathological findings. Sovilla *et al.* (1988) have presented a well documented case of unilateral basal ganglia lesions due to CO with corresponding neurological deficits.

## REFERENCES

Abbott DF (1972) Slow recovery from carbon monoxide poisoning. Postgraduate Medical Journal 48:639-642.

Ali-Cherif A., Royere ML, Gosset A et al. (1984) Troubles du Comportement et de l'Activité Mental après Intoxication Oxycarbonée. Rev. Neurol. 140: 401-405.

Barret L, Danel V, Faure J (1985) Carbon Monoxide Poisoning, A Diagnosis Frequently overlooked. Clinical Toxicology 23: 309-313.

Beard RR, Wertheim GA (1967) Behavioral Impairment Associated with Small Doses of Carbon Monoxide. Am. J. Publ. Health 57: 2012-2022.

Binder JW, Roberts JR (1980) carbon monoxide intoxication in children. Clin toxicol 16:287-295.

Bokonjic N (1963) Stagnant anoxia and carbon monoxide poisoning. Electroencephalor Clin Neurophysiol Suppl. 21:1-102.

Bour H, Tutin N, Pasquier P (1967) The central nervous system and carbon monoxide poisoing. In, Bour H, Ledingham I McA (eds) Carbon Monoxide Poisoning, Amsterdam, Elsevier, pp 1-30.

Bour H, Pasquier P, Bertrand-Hardy JM (1966) Le Coma Oxycarbon, l'etude general, clinique biologique et therapeutique de 290 cas. Sem Hop Paris 42: 1839.

Bunnell DE, Horvath SM (1988) Interactive effects of physical work and carbon monoxide on cognitive task performance. Aviat Space environment Med 59:1133-1138.

Castel Y, Roche J, Le Fur JM et al (1974) Intoxication oxycarbonée chez l'enfant. Cahiers de Medecine 15:795-798.

Choi IS (1983) Delayed Neurologic Sequelae in Carbon Monoxide Intoxication. Arch Neurol 40: 433-35.

DiMarco AT (1988) Carbon monoxide poisoning presenting as polycythemia NEJM 319:874 (letter)

Dorcus RM, Weigand GE (1929) The effect of exhaust gas on performance of certain psychological tests. J Gen Psychol 2:73-76.

Dordain G, Delaporte P, Gray F (1988) Un état confusionnel d'évolution progressive. Rev Neurol (Paris) 144:49-55.

Eltorai IM, Strauss MB, Hart GB et al (1984) Carbon Monoxide Neural Toxicity with Report of A Clinical Case.
Arizone Med 41: 729-34.

Ferguson LS, Burke MJ, Choromokos EA (1985) Carbon Monoxide Retinopathy.
Arch Ophthalmol 103: 66-67.

Funatsu K, Yamada S, Takamuki K, et al (1985) Sauerstoff-Überduckbehandlung der Nachkrankheit bei Kohlenoxidvergiftung.
DMW 110: 140-43.

Garland H, Pearce J (1967) Neurological complications of carbon monoxide poisoning.
Quart J Med 34:445-455.

Garrel S, Perret J, Pellat J et al (1970) Syndrome neuropsychiatrique d'allure frontale, complication post-intervallare d'une intoxication oxycarboneé.
Rev Neurol (Paris) 122:445-447.

Gemelli F, Cattani R (1985) Carbon monoxide poisoning in childhood.
British Medical Journal 291: 1197

Gliner JA, Horvath SM, Mihevic PM (1983) Carbon monoxide and human performance in a single and dual task methodology.
Aviat Space Environ Med 54: 714-7.

Groll-Knapp E, Wagner H, Hauck H et al (1972) Effects of low carbon monoxide concentrations on vigilance and computer-analysed brain potentials.
Staub-Reinhalt Luft 32:64-68.

Halperin MH, McFarland RA, Niven JI et al (1959) The time course of the effects of carbon monoxide on visual thresholds.
J Physiol 146:583-593.

Heckerling PS, Leikin JB, Maturen A (1988) Occult Carbon Monoxide Poisoning: Validation of a Prediction Model.
The American J of Med 84:251-7.

Horvath SM, Dahms TE (1971) Carbon monoxide and human vigilance.
Arch Environ Health 23:343-347.

Kelley JS, Sophocleus GJ (1978) Retinal hemorrhages in acute carbon monoxide poisoning.
JAMA 23 ;1515-1517.

Kirkpatrick JN (1987) Occult Carbon Monoxide Poisoning.
The Western Journal of Medicine 146: 52-6.

Koyabashi K, Isaki K, Fukutani Y et al (1984) CT findings of the interval form of carbon monoxide poisoning compared with neuropathological findings.
Europ Neurol 23:34-43.

Krüger PD, Zorn O, Portheine F (1960) Probleme akuter und chronischer Kohlenoxyd-Vergiftüngen.
Arch Gewerbehygiene 18:1-21.

Littlejohn C (1989) Carbon monoxide inhalation with engine switched off.
Brit J Hosp Med 41:297 (letter)

McFarland RA, Roughton FJW, Halperin MH et al (1944) The effect of carbon monoxide and altitude on visual thresholds.
J Aviat Med 15:381-394.

Meigs JW, Hughs JPW (1952) Acute carbon monoxide poisoning.
Arch Indust Hyg Occup Med 6:344-356.

Mikulka P, O'Donnell R, Heinig P et al (1973) The effect of carbon monoxide on human performance
Ann NY Acad SCi 174: 409-420.

Min SK (1986) a brain syndrome associated with delayed neuropsychiatric sequelae following acute carbon monoxide intoxication.
Acta Psychiatr Scand 73:80-86.

Myers RAK, Snyder SK, Majerus TC et al (1985) Cutaneous blisters and carbon monoxide poisoning.
Ann Emerg Med 14: 603-6.

Myers RAM, Messier LD (1987) The development of a neuropsychological screening baltery for use in clinical assessment of carbon monoxide-intoxicated patients.
In: Kindwall EP (Ed) Pro Eighth International Cong Hyperbaric Med, San Pedro, California, Best Publishing Co., pp. 286-290.

Nick J, Derobert L, Fournier E et al (1965) Manifestations neuorpsychiatriques post-intervallaires de l'intoxication oxycarboneé aigue (10 cas).
Ann Med Leg 36:208-223.

Olson KR (1984) Carbon monoxide poisoning: mechanisms, presentation and controversies in management.
J Emerg Med 1: 233-43.

Plum F, Posner JB, Hain RF (1962) Delayed neurological deterioration after anoxia.
Arch Intern Med 110: 18-25.

Pulst M, Walshe M, Romero A (1983) Carbon Monoxide Poisoning With Features of Gilles de la tourette's Syndrome.
Arch Neurol 40: 443.

Quillam S (1984) Encephalopathy four weeks after carbon monoxide poisoning.
Lancet ii: 408 (letter).

Ray AM, Rockwell TH (1970) An exploratory study of automobile driving under the influence of low levels of carboxyhemoglobin.
Ann NY Acad SC 174:396-407.

Schaad G, Kleinhanß G, Piekarski C (1983) Zum Einfluß von Kohlenmonoxid in der Atemluft auf die psychophysische Leistungsfähigkeit.
Wehrmed. Mschr. 27:423-30.

Schrot J, Thomas JR, Robertson RF (1984) Temporal Changes in Repeated Acquisition Behavior After Carbon Monoxide Exposure.
Neurobehavioral Toxicology and Teratology 6: 23-28.

Schulte JH (1963) Effects of Mild Carbon Monoxide Intoxication.
Arch Environ Health 7: 524-30.

Sokol JA, Krakowska E (1985) The relationship between exposure duration, carboxyhemoglobin, blood glucose, lactate and pyruvate and severity of intoxication in 39 cases of acute carbon monoxide poisoning in man.
Arch Toxicol 57:196-199.

Stewart RD (1975) The effect of carbon monoxide on humans.
Ann Rev Pharmacology 15: 409-23.

Schwartz A, Hennerici M, Wegener OH (1985) Delayed choreoathetosis following acute carbon monoxide poisoning.
Neurology 35: 98-99.

Smith JS, Brandon S (1973) Morbidity from acute carbon monoxide poisoning at three year follow-up.
Br. Med. J 1: 318-21.

Sovilla JY, Despland PA, Bader M (1988) Lésions cérébrales asymétriques et aphasie sous-corticale au cours d'une intoxication au monoxyde de carbone.
Rev Méd Suisse Romande 108:33-40.

Stonesifer LD, Bone RC, Hiller FC (1980) Thrombotic thrombocytopenic purpura in carbon monoxide poisoning.
Arch Int Med 140:104-108.

Taha AM, Kierstead BS, Cardwell R et al (1989) Acute pancreatitis in carbon monoxide poisoning (abstract)
Undersea Biomed Res 16:20.

Trouton D, Eysenck HJ (1961) The effects of drugs on behavior. In, Eyesenck HJ (ed) Handbook of Abnormal Psychology, New York, Basic books, pp 634-696.

Venning H, Roberton D, Milner AD (1982) Carbon monoxide poisoning in an infant.
BMJ 284:651.

Vollmer EP, King BG, Birren JE et al (1946) The effect of carbon monoxide on three types of performance at simulated altitudes of 10,000 and 15,000 feet.
J Exp Psychol 36:244-251.

Werner B, Bäck W, Akerblom H et al (1985) Two Cases of Acute Carbon Monoxide Poisoning with Delayed Neurological Sequelae after a "Free" Interval.
Clinical Toxicology 23: 249-265.

Wiley RW (1989) Dark adaptation and recovery from light adaptation: smoker's versus non-smokers.
Mil Med 154:427-430.

Yastrebov VE, Kustov VV, Razinkin SM (1987) Effect of a Short-Term Exposure to Carbon Monoxide High Concentrations on Man's Psychophysiological Functions. Kosmicheskaya Biologiya I Aviakosmicheskaya Meditsina 21. 47-50.

Yun Dr, Park BJ, Shin Ys et al (1984) An epidemiological study of the neurological sequelae of acute carbon monoxide poisoning. In: Kindwal EP (Ed): Proc. Eighth Int Cong Hyperbaric Med San Pedro, California, Best Publishing Company, pp. 256-257.

Ziser A, Shupak A, Halpern P, et al (1984) Delayed hyperbaric oxygen treatment for acute carbon monoxide poisoning. Br Med J 289: 960.

# Chapter 8

## DIAGNOSTIC PROCEDURES FOR CARBON MONOXIDE POISONING

### INTRODUCTION

Laboratory investigations are required for the definitive diagnosis of CO poisoning as well as for the prognosis. Various diagnostic procedures are listed in Table 8-1. The selection of appropriate tests is done according to the severity of the poisoning, acuteness, or chronicity, and the facilities available at the initial treatment center. Investigations beyond determining COHb concentration in the blood are done according to the system of the body which is mainly involved and the pre-existing illness of the patient.

TABLE 8-1
DIAGNOSTIC PROCEDURES FOR CARBON MONOXIDE POISONING

1. Determination of CO in the blood
   i. Direct measurement of the COHb levels
   ii. Measurement of the CO released from the blood
   iii. Measurement of the CO content of the exhaled air
2. Arterial blood gases and lactic acid levels
3. Screening tests for drug intoxication and alcohol intoxication
4. Biochemistry
   i. Enzymes: creatine kinase, lactate dehydrogenases, SGOT, SGPT
   ii. Serum glucose
   iii. Serum electrolytes
   iv. Blood lactate
   v. Blood ethanol
5. Complete blood count
6. Urine analysis
7. Pregnancy test
8. Chest X-ray
9. Pulmonary function tests
10. ECG
11. Neuropsychiatric testing
12. EEG
13. CT scan
14. Magnetic Resonance Imaging of the brain

## LABORATORY DIAGNOSIS OF CO POISONING

Various techniques for determining CO content of the blood are listed in Table 8-2.

TABLE 8-2
COMPARISON OF SOME TECHNIQUES FOR THE ANALYSIS OF COHb IN BLOOD

| *Type* | *Method sample* | *Blood volume (ml)* | *Analysis difference* | *Detectable (ml/dl)* | *Reference* |
|---|---|---|---|---|---|
| Gasometric | Van Slyke | 1.0 | 15 | 0.3 | Horvath & Roughton 1942 |
| | Syringe-capillary | 0.5 | 30 | 0.02 | Roughton & Root 1945 |
| Optical | Spectro-photometric | 0.1 | 10 | 0.8 | Small et al 1971 |
| | Infra-red | 2 | 30 | .006 | Coburn et al 1964 |
| | Co-Oximeter | 0.4 | 3 | 0.1 | Maasel et al 1970 |
| Chromato-graphy | Thermal conductivity | 1.0 | 30 | .001 | Ayers et al 1966 |
| | Liquid gas chromatography | 0.25 | 3 | .006 | Dahms & Horvath 1974 |

1. **COHb level measurement.** This is the most commonly used investigation. Measurement is done spectrophotometrically offering an accurate and rapid determination of the patients COHb levels. An oximeter such as CIBA-Corning 2500-CO oximeter determines various selected wavelengths from 520-640 nm and the following hemoglobin derivatives are measured.

Oxyhemoglobin ($O_2Hb$)
Deoxyhemoglobin (HHb)
Carboxyhemoglobin (COHb)
Methemoglobin (MetHb)
Oxygen saturation ($sO_2$) can be determined by the following equation.

$$sO_2 = \frac{O_2\,Hb}{O_2Hb + HHb + COHb + MetHb}$$

Absorption spectra of the hemoglobin derivatives are shown in Fig. 8-1.

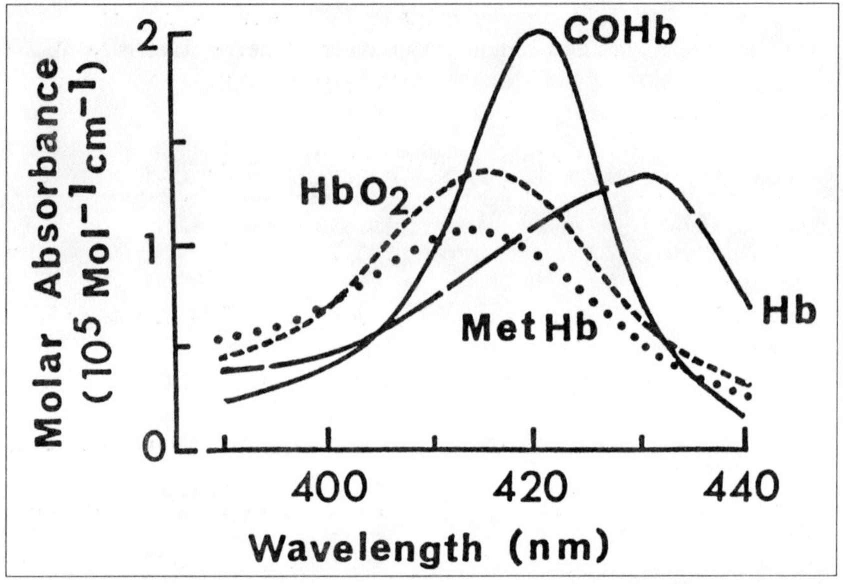


Fig. 8-1: Absorption spectra of various hemoglobin derivatives.

**Determination of CO released from the blood.** Various methods of releasing CO from samples of blood have been reviewed by Heyndrickx *et al.* (1970). CO is then measured by gas chromatography. The amount of CO in the blood sample is calculated from the ratio of the CO content to the full CO capacity of the same sample.

**CO measurement in the exhaled air.** This can be measured by gas chromatography. Hanika (1985) used a bag to collect the exhaled air and determined CO by a flame ionization detector after catalytic hydration to methane. The values are given as ppm (part per million) in the range of 0-500. Equations for estimating COHb from ppm are available (Ringold *et al.* 1962).

**Postmortem determination of COHb.** Three methods have been used to determine COHb satisfactorily in the postmortem blood samples. (Constantino 1986).

1. IL-182 CO-oximeter
2. IL-282 CO-oximeter
3. Head space gas chromatography

If the blood samples are exposed to air, there is a loss of COHb. This can be minimized if the samples are stored in refrigerator (Chace *et al.* 1986).

Losses of up to 60% of the original saturation have been reported for blood kept in uncapped containers at room temperature for two and a half weeks or at 4°C for three weeks. No changes were noticed if the blood was kept in tightly sealed containers (Ocak *et al.* 1985).

**Clinical significance of monitoring blood carboxyhemoglobin.** Fang *et al.* (1986) monitored COHb in 192 patients with acute CO poisoning. They concluded that:

1. Blood COHb>10% has diagnostic significance and COHb> 30% should be considered serious.
2. Clinical manifestations should be primary and COHb secondary when judging the degree of CO poisoning.
3. Treatment should be continued even when COHb levels have returned to normal, if the clinical symptoms are still present.
4. COHb sampling need not be continued when the patient has been away from the toxic environment for more than 8 h.
5. Monitoring of COHb is useful in making a differential diagnosis and in event of death, a definitive diagnosis.
6. Concomitant $H_2S$ poisoning may interfere with the measurement of COHb.

## CHANGES IN THE ARTERIAL BLOOD GASES AND CHEMISTRY

Arterial oxygen tension is usually normal in CO poisoning but oxygen saturation expressed as a percentage is decreased. Some laboratories calculate oxygen saturation rather than measure it with an oximeter. A significant difference between the measured percentage of $O_2Hb$ and calculated $O_2Hb$ percentage is an indication for measuring COHb.

Metabolic acidosis is a feature of CO poisoning and along with enzymes (LDH, CPK, SGOT, and SGPT) is a more sensitive indicator of the severity of poisoning than CO (Olsky and Wood 1983). Increased levels of lactate glucose are influenced by the duration of exposure and are more pronounced after prolonged acute exposure than after short exposure (Sokal 1985). Hyperglycemia may occur due to hormonal stress response.

Blood ethanol levels should be determined in patients with smoke inhalation injury. In a post-mortem study of victims of smoke inhalation injury with high COHb levels, 80% of the adults but none of the children, were found to have blood levels of ethanol above legal limits (Barillo *et al.* 1986).

Myers and Britten (1989) have reviewed their 5-year experience with 247 patients to determine the relation between pH and COHb levels; a weak correlation (4?-0.3) was discovered. They found a much stronger relation of COHb with psychometric testing. These authors rely more on the psychometric test results as an indicator of the severity of CO poisoning.

## CARDIOPULMONARY INVESTIGATIONS

**ECG.** This is a sensitive test for detecting myocardial damage in CO poisoning (Anderson *et al.* 1967). ECG changes in CO poisoning have been described in Chapter 5. Sinus tachycardia is the most common abnormality.

**Pulmonary function tests.** Impairment of pulmonary diffusion capacity may occur with chronic CO poisoning. Insufficiency of obstructive-restrictive type has also been noted in workers exposed to CO (Trajkovic 1986).

**Chest X-ray.** Pulmonary edema is present in 36% of the patients with CO poisoning and is considered to be due to hypoxia. X-rays of the lungs show a characteristic ground glass appearance. Perihaler haze and intra-alveolar edema may also be present (Ogawa *et al.* 2974). Vomiting in an unconscious patient may lead to aspiration pneumonia. Presence of pulmonary manifestations indicates a poor prognosis (Sonne *et al.* 1974).

## NEUROPSYCHOLOGICAL TESTING

Various psychological tests have been designed for patients with CO poisoning. Six tests of memory, comprehension, spatial orientation, motor speed, fine motor control, and visual coordination have been described by Dolan (1985).

Myers and Messina (1987) have described a neuropsychological screening batter for use in the assessment of CO-intoxicated patients. This consists of six tests:

1. General orientation
2. Digit span
3. Trail making
4. Digit symbols
5. Aphasia
6. Block design

These tests can be administered in an emergency setting by non-psychologists in less than 20 minutes. The test battery has been shown to discriminate effectively and reliably between patients and controls.

**Electroencephalographic changes.** Ogawa *et al.* (1973) reported that 97% of the patients suffering from CO poisoning had EEG abnormalities within 24 hours, in spite of treatment. The various EEG abnormalities noticed in CO poisoning are;

Diffuse abnormalities—continuous theta or delta activity
Low voltage activity accompanied by intervals of spiking or silence
Rhythmic bursts of slow waves

In his classic study, Bokonjic (1963) differentiated the EEG in cases of CO poisoning from that in cases of stagnant anoxia; the voltage was reduced in the former but was never as severe as the flattening in the latter.

**CT changes.** Miura *et al.* (1985) reported abnormalities on CT scan in 23 of the 60 patients with acute CO poisoning. The most common finding seen in 21 patients was symmetrical and diffuse low density lesions in the cerebral white matter. In 18 patients there were bilateral, round low-density lesions in the globus pallidus. The prognosis and recovery correlated with the severity of the cerebral white matter changes and not the size of the globus pallidus lesions. The lesions may be unilateral (Taylor *et al.* 1988). CT scan of a patient with CO poisoning is shown in Figure 8-2.

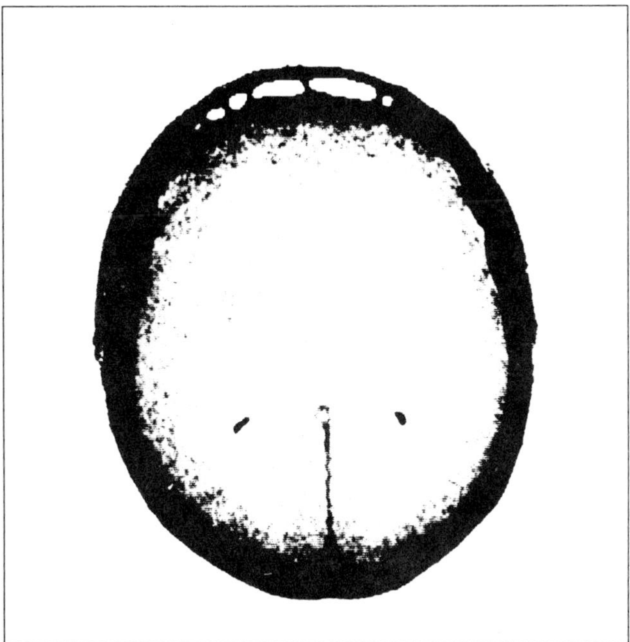

Fig. 8-2: CT scan 5 days after severe CO poisoning. Symmetrical zone of necrosis in the region of globus pallidus bilaterally. This patient recovered completely after oxygen therapy without any residual motor disorder (Reproduced from: Berlit P, Johann A, Bühler B: Akt Neurol 15 (1988) 49-52), by permission of the authors.

Not all patients with CO poisoning have abnormal CT scans (Sawada *et al.* 1980). Delayed CT changes with contrast enhancement associated with slightly elevated COHb level (4.5%) have been reported on the eighth hospital day in a comatose patient. The changes resolved as the patient improved during the subsequent five weeks (Nardizzi 1979).

Fife *et al.* (1989) studied the prognostic value of CT scan in 15 patients with CO poisoning undergoing HBO treatments. CT abnormalities included either basal ganglia infarcts (5 patients) or diffuse cerebral edema (one patient). All patients with positive CT scans had poor outcome in spite of HBO treatment; two of them died, one had parkinsonism and dementia, one had dysphasia, one had severe personality changes, and one became partially blind.

Magnetic resonance imaging (MRI) and single photon emission tomography (SPECT) have not yet been used widely in the diagnosis of CO poisoning. These procedures can provide more detailed information in cases of CO encephalopathy.

# REFERENCES

Anderson R, Allensworth DC, De Groot WJ (1967) Myocardial toxicity from carbon monoxide poisoning.
Ann Int Med 67:1172-1182.

Ayres SM, Criscitello A, Gianelli S Jr (1966) Determination of blood carbon monoxide content by gas chromatography.
J Appl Physiol 21:1368-1370.

Barillo DJ, Rush BE, Good R et al (1986) Is ethanol the unknown toxin in smoke inhalation injury?
Am Surg 52:641-645.

Bokonjic N (1963) Stagnant anoxia and carbon monoxide poisoning.
Electroenceph Clin Neurophysiol Suppl. 21, 1-102.

Chace DH, Goldbaum LR, Lappas NT (1986) Factors Affecting the Loss of Carbon Monoxide from Stored Blood Samples.
Jouranl of Analytical Toxicology 10:181-189.

Coburn RF, Danielson GK, Blakemore WS et al (1964) Carbon monoxide in blood: analytical methods and sources of error.
J Appl Physiol 19:510-515.

Costantino AG, Park J, Caplan YH (1986) Carbon Monoxide Analysis: A Comparison of Two Co-Oximeters and Headspace Gas Chromatography.
J of Analytical Toxicology 10: 190-193.

Cox BD, Whichelow MJ (1985) Carbon monoxide levels in the breath of smokers and nonsmokers: effect of domestic heating systems.
J of Epidemiology and Community Health 39: 75-8.

Dahms TE, Horvath SM (1974) Rapid, accurate, technique for determination of carbon monoxide in blood.
Clin Chem 20:533-537.

Davis SM, Levy RC (1984) High carboxyhemoglobin level without acute or chronic finding.
J Emergency Med 1:539-542.

Fang GC, Xu GH, Wang FM (1986) Clinical significance of monitoring blood carboxyhemoglobin. J Hyperbaric Med 1:233-238.

Fife CE, Sallee DS, Gray L, Piantadosi CA (1989) Prognostic value of CT scan after severe CO poisoning (abstract).
Undersea biomed Res 16:19.

Hanika G (1985) Gaschromatographische Messung des Kohlenmonoxids in der Exhalations-luft im ppm Bereich.
Z. gesamte Hyg 31: 574-77.

Heyndrickx A, Scheiris CH, Vercruysse A et al (1970) Gas chromatographic determination of carbon monoxide in blood and the hyperbaric oxygen treatment in carbon monoxide poisoning cases.
J Pharmac de Belgique: 247-258.

Horvath SM, Roughton FJW (1942) Improvements in the gasometric estimation of carbon monoxide in blood.
J Biol Chem 144:747-755.

Huber JA (1984) Do awake patients with high carboxyhemoglobin levels need hyperbaric oxygen?
J Emerg Med 1:555-556.

Mass AHJ, Hamelink ML, Deleeuw RJM (1970) An evaluation of the spectrophotometric estimation of HbO2, and Hb in blood with CO-oxumeter IL 182.
Clin Chim Acta 29:303-309.

Miura T, Mitomo M, Kawai R et al (1985) CT of the Brain in Acute Carbon Monoxide Intoxication: Çharacteristic Features and Prognosis.
AJNR 6:739-42

Myers RAM, Britten JS (1989) Are blood gases of value in treatment decisions for carbon monoxide poisoning?
Crit Care Med 17:139-142.

Myers RAM, Messier LD (1987) the development of a new neuropsychiatric screening battery for use in clinical assessment of carbon monoxide-intoxicated patients. In, Kindwall EP (ed): Proceedings of the Eighth International Congress of Hyperbaric Medicine, San Pedro, California, Best Publishing company, pp 286-290.

Nardizzi LR (1979) Computerized tomographic correlate of carbon monoxide poisoning.
Arch Neurol 36:38-39.

Nicholson NA, Karabus CD, Hartley PS (1988) Reduction of neonatal blood eliminates the error induced by oxygen saturation in the spectrophotometric measurement of carboxyhemoglobin.
Ann Clin Biochem 25:237-241.

Ocak A, Valentour JC, Blanke RV (1985) The Effects of Storage Conditions on the Stability of Carbon Monoxide in Postmorten Blood.
J of Analytical Toxicology 9:202-206.

Ogawa M, Minami T, Katsurada K et al (1973) Early Electroencephalographic Change Following Acute Carbon Monoxide Poisoning in Relation to Cerebral Metabolism.
Medical Journal of Osaka University 24: 85-90.

Ogawa M, Katsurada K, Sugimoto T et al (1974) Pulmonary edema in acute carbon monoxide poisoning.
Int Arch Arbeitsmed 33: 131-138.

Ringold A, Goldsmith JR, Helwig HL (1952) Estimating recent CO exposure.
Arch Environ Health 5: 308-18.

Roughton FJW, Root WS (1945) The estimation of small amounts of carbon monoxide in blood. J Biol Chem 160:123-133.

Sawada Y, Takahashi M, Ohashi N et al (1980) computerized tomography as an indication of long-term outcome after acute carbon monoxide poisoning.
Lancet i:783-784.

Small KA, Radford EP, Frazier JM et al (1971) A rapid method for simultaneous measurement of carboxy- and methemoglobin in blood.
J Applied Physiol 31: 154-160.

Sokol JA (1985) the effect of exposure duration on the blood level of glucose pyruvate and lactate in acute carbon monoxide intoxication in man.
J Appl Toxicol 5:395-397.

Sone S, Higoshihari T, Kotake T et al (1974) Pulmonary manifestations in acute carbon monoxide poisoning.
Am J Roent 120:865-871.

Taylor R, Holgate RC (1988) Carbon Monoxide Poisoning: Asymmetric and Unilateral Changes on CT.
AJNR ).975-977.

Trajkovic M (1984) Study of pulmonary ventilation function in workers exposed to carbon monoxide. (in russian)
Srpki Arhiv Za Selokupno Lekarstvo 112:909-917.

# Chapter 9

# GENERAL PRINCIPLES OF MANAGEMENT OF CARBON MONOXIDE POISONING

## GUIDELINES FOR MANAGEMENT OF CO POISONING

General guidelines for the management of carbon monoxide (CO) poisoning are shown in Table 9-1. The objects of treatment of CO poisoning are as follows:

1. To hasten the elimination of CO from the body.
2. To counteract hypoxia and its effects on the tissues of the body.
3. To counteract direct tissue toxicity of carbon monoxide.

TABLE 9-1
GUIDELINES FOR THE MANAGEMENT OF CARBON MONOXIDE POISONING

- Remove patient from the site of exposure
- Administer oxygen immediately. If possible take a blood sample for COHb before this.
- Use endotracheal intubation in comatose patients to facilitate ventilation
- Remove the patient to a hyperbaric facility when indicated
- Keep the patient calm and avoid physical exertion by the patient
- Management of complication: e.g. electrolyte imbalance, cardiac arrythmias.

## OVERVIEW OF VARIOUS METHODS OF TREATMENT

An overview of the various methods used in treatment of carbon monoxide poisoning and its sequelae is shown in Table 9-2. Obsolete treatments are listed in Table 9-3.

## RESUSCITATION IN ACUTE SEVERE CO POISONING

The most important first step in the case of acute severe CO poisoning is removal of the patient from CO environments. CO slowly dissociates from the hemoglobin and is eliminated. The following initial steps are taken for a comatose patient:

TABLE 9-2
OVERVIEW OF VARIOUS METHODS USED TO TREAT CARBON MONOXIDE POISONING

| *Aim of treatment* | *Method* | *Comments* |
|---|---|---|
| Elimination of CO | Oxygen therapy both NBO and HBO | Mass action of oxygen in blood to displace CO. HBO is more effective than NBO |
| Reversal of tissue hypoxia | Oxygenation: NBO and HBO | HBO is more effective than NBO. High saturation of plasma with oxygen in spite of high COHb levels. |
| Cellular protection | $Mg^{2+}$ therapy | Cell stabilizing effect of $Mg^{2+}$ |
| *Management of sequelae* | | |
| a. Cerebral edema | 1. Mannitol<br>2. Corticosteroids<br>3. HBO | Short-term dehydrating effect<br>Effectiveness is controversial.<br>Effective. Cerebral vasoconstriction reduces edema. |
| b. Pulmonary edema | 1. Diuretics<br>2. HBO | |
| c. Cardiac complications | 1. Cardiac glycosides<br>2. Anti-arrythmic drugs | |
| d. Pneumonia | Antibiotics | |
| e. Acidosis | Sodium bicarbonate. | Avoid full correction of acidosis. Watch for arrythmia if hypokalemia is present. |
| f. Vitamin deficiency | Vitamin B-complex | Replacement therapy |
| | Vitamin K | To prevent hemorrhages due to liver damage |

HBO- Hyperbaric Oxygen, NBO- Normobaric Oxygen

TABLE 9-3
PAST THERAPIES FOR CARBON MONOXIDE POISONING WHICH HAVE NOT PROVEN TO BE EFFECTIVE

| *Therapy* | *Author* | *Comments* |
|---|---|---|
| 1. Intravenous procaine | Aymes et al 1950 | Was used to treat anoxia of CO poisoning. Amyes (1989) suggests that it be replaced by lidocaine which has been shown to facilitate return of neural function after cerebral ischemia in experimental animals (Evans *et al.* 1989). |
| 2. Cytochrome c | Gros 1956, Stelter 1953 | Used for activating cytochrome oxidase. Not used in humans |
| 3. Methylene blue | Bassaberger 1934 | Not an antidote but a synergist (Haggard & Greenberg 1933). |
| 4. Succinic acid | Gershon et al.1961 | |
| 5. Persentin | Frimmer & Hegner 1963 | |
| 6. Iron and cobalt preparations | Paulet 1963 | |
| 7. Ascorbic acid | Prisco 1941 | |
| 8. Hydrogen-peroxide infusions | Bentolila et al 1973 | Reduces COHb content in experimental animals. No human experience. Danger of gas embolism. |
| 9. Ultraviolet radiation | Koza et al. 1930 | Supposedly facilitates the dissociation of COHb from RBCs during transit through skin capillaries. Extracorporeal exposure of blood to ultraviolet light reduces the mortality in CO-poisoned animals from 100% to 20%. Exposure of animals to ultraviolet light has been shown to be ineffective in lowering COHb (Estler 1935). |

1. Keep the airway patent and suction and suction aspirated material from the trachea.
2. Endotracheal intubation
3. Central venous catheter placement and verification by X-ray.
4. Gastric intubation via nostrils and stomach lavage.
5. Urinary catheter and hourly urine output measurements.

## OXYGEN THERAPY

Inhalation of 100% oxygen is an important initial step in treatment. It facilitates the elimination of CO by mass action. Further management of the patient can be planned with the use of a triage decision chart shown in Figure 9-1.

Fig. 9-1: Carbon Monoxide—triage decision chart
*(Reproduced from Kindwall E, Goldmann RW: Hyperbaric Medicine Procedure, St. Luke's Medical Center, Milwaukee, Wisconsin, 1988. By permission.)*

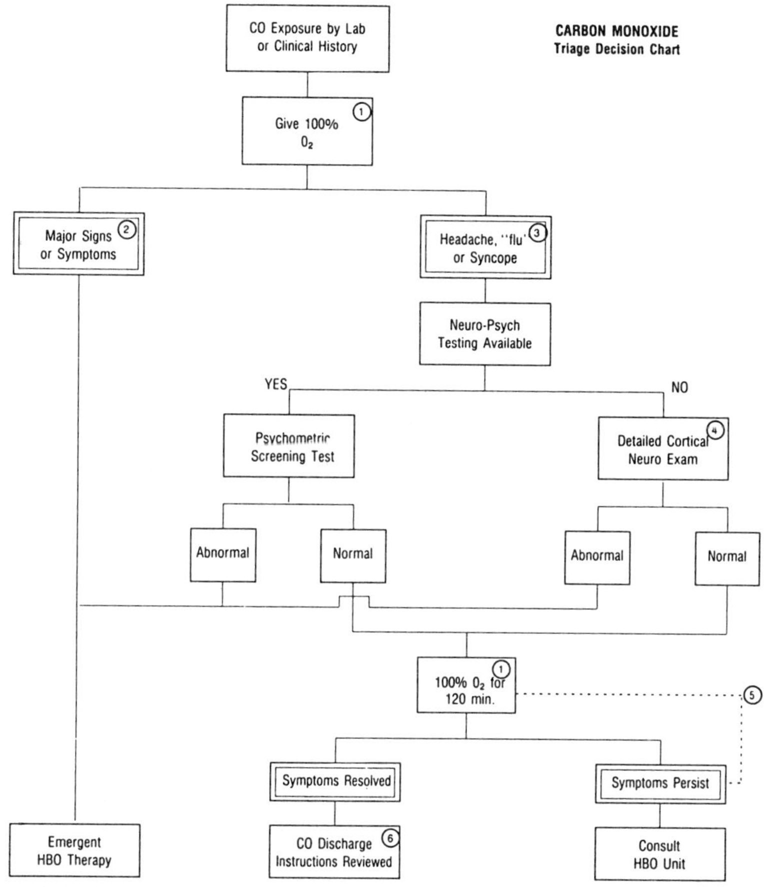


1. $O_2$ should be delivered by tight fitting system such as Scott mask, anesthesia mask, endotracheal tube or CPAP mask.
2. Major signs and symptoms include: abnormal EKG, metabolic acidosis, lab or clinical findings of pregnancy, chest pain, confusion, disorientation, personality change, lethargy or drug overdose with mental status change.
3. Headache may be severe and mimic intracranial hemmorhage in severity. Any suggestion of post syncopal neurological dysfunction is a major symptom.
4. Detailed cortical neurological examination should include: general orientation, phone number, address, date of birth, serial 7's, digit span, forward and backward spelling of three and four letter words and short term memory.
5. OPTION: One repeat two hour $O_2$ cycle (4 hours of surface $O_2$ total) is permissible. If symptoms persist beyond this point, consultation and possible referral is indicated.
6. Follow up of CO exposed patients is critical. Recurrent or indolent symptoms or family observation of abnormalities should be re-evaluated emergently.

SLMC/HBO

Oxygen inhalation reduces the half-life of CO as shown in Table 2-5 (Chapter 2). At atmospheric pressure in fresh air, the circulating half-life of CO is 5 hr 20 min. This time decreases to 23 min with hyperbaric oxygen at a pressure of two atmospheres absolute (2 ATA) using 100% oxygen. Techniques of oxygen administration are described in detail elsewhere (Jain 1989). It is essential to use a tight-fitting mask.

The most effective form of oxygen therapy is HBO which is described in more detail in Chapter 10.

## MANAGEMENT OF SEQUELAE

Various methods for treating the sequelae of CO poisoning are as follows:

1. Cerebral edema. Use of steroids for relieving cerebral edema is controversial. Mannitol gives only short term relief. HBO therapy, however, is itself effective against cerebral edema.

2. Cellular protection. $Mg^{2+}$ exerts a protective and stabilizing effect on cellular membranes. I recommend $Mg^{2+}$ in daily dosage of 20-30 m mol as an adjunctive measure for cellular protection (Jain 1990). The use of allopurinol combined with N-acetylcysteine has been recommended by Howard *et al.* (1987).

3. Fluid and electrolyte imbalance should be corrected. Overhydration, which aggravates cerebral edema and pulmonary complications, should be avoided. Acidosis should not be corrected pharmacologically because slight acidosis aids in the delivery of oxygen to the tissues by shifting the oxygen dissociation curve to the right. A pH of 7.15 is ideal. Caution should be exercised in the use of sodium bicarbonate, because in the presence of hypokalemia, which is a frequent occurrence, cardiac arrythmias may be induced.

4. Pulmonary edema. It is a common finding in smoke inhalation injuries. Treatment of acute respiratory distress (ARD) syndrome requires intensive respiratory care where supplemental oxygen inhalation therapy has a useful role.

5. Cardiac arrythmias. Cardiac arrythmias are a common complication of CO poisoning. They may subside with reversal of tissue hypoxia but anti-arrythmic drugs may be required.

6. Hyperthermia. This may occur as a complication of CO poisoning and should be treated intensively to reduce the tissue oxygen requirements. Dantrolene sodium, which is used for the treatment of malignant neuroleptic syndrome, has been used successfully in a patient with CO poisoning and hyperthermia associated with excessive rigidity (Ten Holter and Schellens 1988).

Kaltwasser *et al.* (1977) investigated the mortality from CO poisoning in 187 patients in relation to body temperature; it was 22% in normothermic but 36% in hyperthermic or hypothermic patients. Following further animal experiments, these authors stated that it is not advisable to maintain hypothermia during CO poisoning.

Sluijter (1963) found hypothermia to be useful in the management of CO poisoning when used in conjunction with HBO. Pierce *et al.* (1972), on the other hand, found hypothermia to be inferior to HBO in the treatment of experimental CO poisoning in animals. The explanation for this that CO has a greater affinity for Hb at reduced temperature; 400 to 600 times more than oxygen at 15° C versus 200 to 300 times at 372C (Killick 1940). Hypothermia in conjunction with 100% oxygen inhalation has been recommended for the management of abnormal motor activity due to CO poisoning if no hyperbaric chamber is available (Boutros and Hoyt 1976).

## MISCELLANEOUS METHODS OF UNPROVEN VALUE

The use of 5% $CO_2$ has been advocated on the assumption that it increases the ventilatory response and thus facilitates CO elimination. It can be used if the patient is on intermittent positive pressure breathing and the blood gases are controlled to avoid hypercarbia (Unsworth 1974).

Carbogen (95% oxygen + 5% $CO_2$) has been recommended to facilitate the release of oxygen from COHb (Haggard 1920, Hasimoto 1938, Henderson and Haggard 1922, Martland 1934). Carbogen does not normalize the pH of blood, nor does reduce the blood lactate any faster than oxygen. Schwerma *et al.* (1948) exposed dogs to CO and found no increase in survival rate in groups who breathed 97% oxygen plus 3% $CO_2$ as compared to animals who breathed 100% oxygen. Due to danger of acidosis and increased oxygen utilization, carbogen has not been recommended for CO poisoning. If barbiturate poisoning coexists with CO poisoning, carbogen is contraindicated as it leads to $CO_2$ retention.

Blood exchange is an old method of treating CO poisoning. Blood letting has been used for this purpose (Koch 1939) but the value is questionable. In special cases, azotemia due to renal failure resulting from CO poisoning may be treated by hemodialysis (Bowen 1960) or peritoneal dialysis.

Extracorporeal oxygenation has been used in patients who were comatose and in poor general condition following CO poisoning (Radushkevich *et al.* 1976). The immediate effects were reported to be favorable.

Exchange transfusions have been used in severe CO poisoning but the value is questionable. Yee and Brandon (1983) reported a case where this was used in combination with HBO and it is difficult to judge which of the two therapies was the effective agent.

## MANAGEMENT OF ASSOCIATED INJURIES

Assessment for other injuries including those of cervical spine, burns, fractures, or pneumothorax should be made.

Smoke inhalation involves multiple toxicities and pulmonary insufficiency, as well as thermal and chemical injury. CO intoxication is the most immediate life-threatening disorder in such cases. Role of HBO in the management of smoke inhalation injury is described in Chapter 10.

## CRITERIA FOR ADMISSION TO HOSPITAL

The decision for admission of a patient with CO poisoning is based on clinical assessment, individual susceptibility, response to initial therapy and COHb levels. Patients can be treated and discharged from the emergency department. The following should be admitted:

1. Pregnant patients with hemoglobin > 10%.
2. Patients with cardiovascular disease and symptom of chest pain or abnormal ECG.
3. Patient with history of unconsciousness or abnormalities on neuropsychological examination.
4. Patients with metabolic acidosis.

## MANAGEMENT OF CHRONIC OCCULT CO POISONING

The most important aspect of the management of chronic CO poisoning is the detection and removal of the cause such as a defective gas oven or heating system with leakage of CO. Those who are exposed to high concentrations of CO at work such as on streets with heavy traffic, should be removed to fresh air. This is usually sufficient to lower the blood COHb levels.

Chronic smokers with high COHb levels should be advised to stop smoking. No method of lowering COHb or treatment of symptoms is effective without smoking cessation. Measures for smoking cessation are described in Chapter 11.

## REFERENCES

Amyes EW, Ray JW, Brockman NW (1950) Carbon monoxide anoxia; intravenous administration of procaine hydrochloride in the treatment of acute and chronic effects. JAMA 142:1054-1058.

Amyes EW (1989) Lidocaine in treatment of hypoxia (letter). J Neurosurg 71:153.

Bentolila P, Tran G, Olive G (1973) Essai de traitment de l'intoxication oxycarbonée par perfusion de solutions diluées de paroxyde d'hysrogene. Resultats obtenus chez le lapin. Thérapie 28:1043-1049.

Boutros AR, Hoyt JL (1976) Management of carbon monoxide poisoning in the absence of a hyperbaric chamber. Crit Care Med 4:144-147.

Bowen DAL (1960) Acute renal failure in carbon monoxide poisoning.
J Forensic Med 7:78-87.

Bussaberger RA (1934) Glutathione and methylene blue in carbon monoxide poisoning.
Proc Soc Exp Biol Med 31:598-600.

Estler W (1935) Experimentelle Untersuchungen über die Anwendung der Bestrahlung mit ultra-voilettem Licht zur Behandlung der Kohlenoxidvergiftung.
Arch Hyg Bakt 115:152-157.

Evans DE, Catron PW, McDermott JJ et al (1989) Effect of lidocaine after experimental cerebral ischemia induced by air embolism.
J Neurosurg 70:97-102.

Frimmer M, Hegner D (1963) Beeinflussung der Kohlenmonoxidvergiftung der Maus durch Persantin.
Klin Wochenschr 41:1165-1166.

Gershon S, Trethewie ER, Crawford M (1961) The use of succinic acid in the treatment of acute carbon monoxide poisoning.
Arch Internat Pharmacodynam Thér 134:16-27.

Gros JF, Leandri P (1956) Traitment de l'intoxication oxycarbonée par le cytochrome.
Presse Méd 64:1356-1357.

Haggard HW, Greenberg La (1933) Methylene blue: A synergist, not an antidote, for carbon monoxide.
JAMA 100:2001-2003.

Howard RJM, Balake DR, Pall H et al (1987) Allopurinol/N-Acetylcysteine for carbon monoxide poisoning.
Lancet ii:628-629.

Jain KK (1989) Oxygen in Physiology and Medicine.
Springfield, Illinois, Charles C. Thomas

Jain KK (1990) Cerebral Insufficiency. Chicago, year Book Medical Publishers.

Kaltwasser K, Kleinau H, Pankow D et al (1977) Hypothermie und Kohlenmonoxid Toxität.
Z Exp Chirug 10:45-51.

Killick EM (1940) Carbon monoxide anoxemia.
Physiol Rev 20:313.

Koch KG (1939) Die Behandlung der akuten schweren Kohlenmonoxidvergiftung mit Bluttransfusionen.
Münch Med Wochenschr 86:126-128.

Koza F (1930) Die Kohlenmonoxidvergiftung und deren neuartige Therapie mit Bestrahlung.
Med Klinik 26:422-425.

Paulet G (1963) Etude de l'action du cobalt dans l'intoxication par l'oxide de carbone.
Arch Mal Profess Méd Trav Séc Soc 24: 426-429.

Pierce EC, Zacharias A, Alday JM et al. (1972) Carbon Monoxide Poisoning: experimental hypothermic and hyperbaric studies.
Surgery 72:229-237.

Prisco L (1941) Sull'azione dell'acido ascorbico nell'intossicazione ossicarbonica.
Boll Soc Ital Biol Sperim 16:641-644.

Radushevich VP, Koroteeva EL (1976) Parallel blood circulation with oxygenation of blood in severe poisoning with carbon monoxide fumes.
Vestn Khir 116: 131-134.

Schwerma H, Ivy AC, Friedman H et al (1948) Study of resuscitation from justalethal effects of exposure to carbon monoxide.
Occupat Med 5:24-48.

Sluitjer ME (1963) The treatment of carbon monoxide poisoning by administration of oxygen at high pressure.
Proc Roy Soc Med 56: 1002-1008.

Stelter R (1953) Zur Therapie der akuten Kohlenmonoxidvergiftung mit Cytochtrome C.
Die Medizinische WSchr:351-352.

Unsworth IP (1974) Acute carbon monoxide poisoning.
Anesth Inten Care 4: 329-332.

# Chapter 10

## TREATMENT OF CO POISONING BY HYPERBARIC OXYGENATION

Hyperbaric oxygenation (HBO) is administration of oxygen under pressure greater than that found at sea level. The most widely used terminology to denote the pressure is atmospheres absolute or ATA. The physics and physiology of HBO is explained in detail elsewhere (Jain 1990).

When breathing ambient air, the blood contains about 20% of oxygen, 19.7% combined with hemoglobin and 0.3% dissolved in plasma. When 100% oxygen is breathed at 1 ATA, the hemoglobin is saturated (20.1% $O_2$), but oxygen contained in plasma rises to 1.88 vol %. At 3 ATA, oxygen dissolved in plasma is 6 vol % with a $pO_2$ of 2193 mm Hg (Fig. 10-1). This volume of oxygen is sufficient to sustain life in the complete absence of functional hemoglobin.

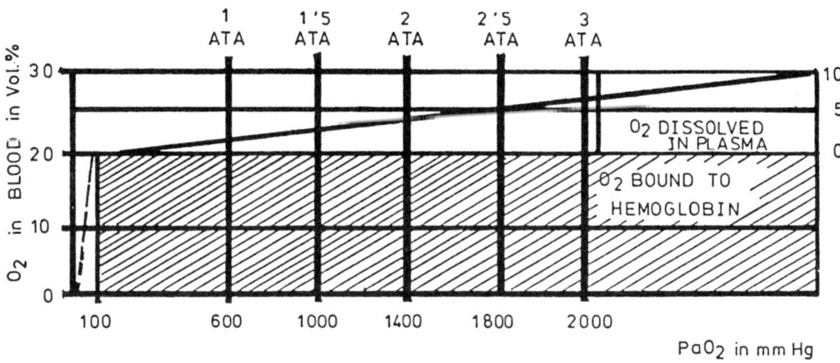


Fig. 10-1: Arterial oxygen tension ($PaO_2$) increase and volume of oxygen dissolved in blood in relation to 100% oxygen breathed at various atmospheric pressures (ATA).

### RATIONALE OF THE USE OF HBO IN CO POISONING

HBO accomplishes the following therapeutic goals in CO poisoning:

1. Immediate saturation of plasma with enough oxygen to sustain life and to counteract tissue hypoxia in spite of high levels of COHb.

2. It causes a rapid reduction of CO in the blood by mass action of $O_2$. In the equation

$$HbO_2 + CO = HbCO + O_2,$$

an increase in either oxygen or CO results in a comparable increase in corresponding compound with hemoglobin.

3. It assists in driving CO away from cytochrome oxidase and in restoration of function. Increase in oxygen tension in plasma and not simply an increase in dissolved oxygen is responsible for the efficacy of HBO.

4. HBO reduces cerebral edema (Sukoff *et al.* 1968).

## EXPERIMENTAL EVIDENCE

The results of some experimental use of HBO in CO poisoning are shown in Table 10-1.

TABLE 10-1
EXPERIMENTAL STUDIES OF THE EFFECT OF HYPERBARIC OXYGENATION ON CARBON MONOXIDE POISONING

| *Author and year* | *Experimental subjects* | *Mode of oxygen therapy* | *Results* |
|---|---|---|---|
| End and Long (1942) | Dogs and guinea pigs | HBO-3ATA, 100% oxygen | HBO more effective than normobaric oxygen in eliminating CO from the body. |
| Pace et al (1950) | Human volunteers | HBO-2ATA | Rate of diminution of CO accelerated |
| Ogawa et al (1974) | Dogs | HBO | Hemoconcentration (20% decrease of blood volume reversed by HBO). |
| Koyayama et al (1976) | Dogs | Half of the animals treated by conventional methods and the other half by HBO(2ATA) | COHb determination and biochemical studies showed that HBO was more effective than the conventional methods. |
| Sasaki (1975) | Dogs | HBO | Acceleration of the half-clearance time of COHb, Proposed procedure for HBO therapy based on it.<br>1. For severe CO poisoning, HBO at 2.8 ATA for 20 min followed by 1.9 ATA for 57 min.<br>2. For moderate CO poisoning, 2,8 ATA oxygen for 20 min followed by 1.9 ATA for 46 min.<br>3. For light CO poisoning, 2.8 ATA oxygen for 20 min followed by 1.9 ATA for 30 min. |

Haldane (1895) exposed a mouse in a jar containing 2 atmosphere of oxygen and one atmosphere of CO and failed to produce clinical signs of CO poisoning. he recommended that this hyperoxic technique be used in the treatment of CO poisoning. End and Long (1942) were the first to show that HBO was more effective than normobaric oxygen in treating CO poisoning in experimental animals.

Valois and Schadé (1967) studied the electrophysiology of histotoxic anoxia with CO in experimental animals. The hippocampus was found to be particularly vulnerable. The hippocampal EEG recovery time was found to be significantly shorter than the neocortical EE recovery time. At 3 ATA of 100% oxygen, these values became significantly smaller.

Araki *et al.* (1988) performed non-invasive optical monitoring of COHb with visible reflectance spectroscopy. They found that there is a close correlation between optical signals and COHb content determined from blood samples. They were able to show that HBO treatment markedly accelerates dissociation of COHb.

## TECHNIQUES OF HBO ADMINISTRATION

The treatments may be carried out in a monoplace or a multiplace hyperbaric chamber. A large chamber with intensive care facilities is preferable in case of a critically ill patient. The inside view of a multiplace hyperbaric chamber is shown in Figure 10-2. There are some variations in the treatment schedules. The general procedure at the Maryland Institute for Emergency Medical Services, Baltimore, is 46 min of 100% oxygen at 3 ATA followed by 2 hours of further treatment at 2 ATA or until the COHb is less than 10% (Myers *et al.* 1979).

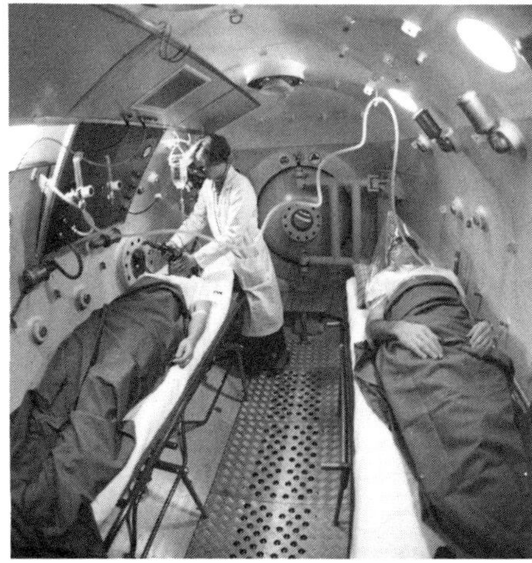

Fig. 10-2: Inside view of a hyperbaric chamber.

## CLINICAL REPORTS OF THE USE OF HBO IN CO POISONING

Sharp and Smith (1960) were the first to treat human patients with HBO for CO poisoning. The first two cases recovered completely. Some of the subsequently reported series in literature are summarized in Table 10-2.

TABLE 10-2
CLINICAL APPLICATIONS OF HBO IN CO POISONING

| *Author and year* | *No. of patients* | *Pressure* | *Results* |
|---|---|---|---|
| Smith et al (1962) | 22 | 2 ATA | All recovered |
| Sluitjer (1963, 1967) | 40 | 3 ATA | Group I: conscious or drowsy. 21 patients. Excellent results. Group II: comatose with neurological abnormalities. 10 patients. Two died, 7 recovered fully and one had sever neurological sequelae. Group III: Attempted suicide with combination of CO and barbiturates. 9 patients. Cardiorespiratory depression mainly with little localizing neurological signs. All recovered completely. |
| Goulon et al (1969) | 302 | 2 ATA | Mortality when treatment started before 6 h was 13.5%, and when after 6 h 30.1%. |
| Heyndrickx (1970) | 11 | 3 ATA | Clinical improvement more than in other 11 patients treated by NBO. |
| Kienlen et al (1974) | 370 | 2-3 ATA | 93.7% of the patients recovered. |
| Adamiec et al (1975) | 44 | 2.5 ATA | 80% showed good recovery. |
| Yun and Cho (1983) | 2242 | ? | 98.2% recovered. |
| Mathieu et al (1985) | 203 | ? | Mortality 1.7%; Incidence of secondary syndromes, 4%: rest recovered. |
| Norkool (1985) | 115 | ? | 88% recovered fully. |
| Colignon and Lamy (1986) | 1118 | 3 ATA | 0.5% died in emergency room; 3.3% admitted to ICU; rest 96.2% had minor symptoms and recovered completely. |
| Tirpitz & Baykara (1988) | 276 | 2-2.5 ATA | 4 death. Rest recovered. Many treated and released to home the same day. |
| Sloan et al (1989) | 297 | 3 ATA | Extremely ill patients with mortality of 6%. Rest recovered. See text. |

It is difficult to compare the results of different studies because the patient conditions differed widely and the HBO technique used also varied. The overall results were favorable. Some patients in critically ill condition died from other complications and in some other cases the HBO therapy was instituted too late to be life-saving. Nevertheless, HBO therapy may still be useful even if applied late as has been shown by Ziser *et al.* (1984).

In those cases, where HBO is combined with other treatment modalities, it may be difficult to assess the role of HBO. Radushkevich and Korteva (1976) reviewed 141 patients with acute CO poisoning. In 110 of these cases, the poisoning was severe and the therapy was delayed. HBO was combined with extracorporeal circulation. Fourteen patients died due to pneumonia and meningoencephalitis but the authors concluded that HBO was beneficial in these cases.

Several authors have presented one or two cases of CO poisoning but they are instructive. Neubauer (1979) reported a case seen 13 days after exposure to CO. The patient was comatose and decerebrate with cerebral edema shown on CT scan and a grossly abnormal EEG. The patient woke up during the first HBO session and required further treatments. After twenty-third HBO session the patient could walk without assistance. The mental symptoms improved as well. Winter and Sharin (1970) presented a case of CO poisoning with coma from which the patient recovered after HBO.

Welsh and De Travello (1980) reported two cases of CO poisoning. One of these had been comatose for 33 hr and did not respond to HBO. The other case with milder CO poisoning made a considerable recovery after HBO treatment. Yee and Brandon (1983) treated a semicomatose patient with CO poisoning by simultaneous use of HBO, exchange transfusions, and steroid therapy. The patient made a complete recovery.

Rioux and Myers (1989) presented two cases of CO poisoning resulting from exposure to methylene chloride in paint. Both were treated successfully with HBO and recovered. Youn *et al.* (1989) treated 12 patients with CO poisoning resulting from a single incident of methylene chloride spill. There was a delay of several hours between the incident and treatment. Diagnosis of CO poisoning was established by serial neuropsychological tests and all of these had an early resolution of symptoms without any sequelae.

In 1985, the North American Hyperbaric Center (affiliated with Bronx Municipal Hospital) carried out HBO treatments on 81 cases of severe CO poisoning, the results of which were reported recently (Hyperbaric Center Advisory committee, EMS, New York City 1988). The patients had lost consciousness and nineteen of these were still comatose at the time of admission. Following HBO treatment (single session of 46 minutes at 3 ATA), 16 of these started to wake up. Four of the children did not survive. There were no complications in the pregnant women. The program of HBO treatments was considered to be a success.

## CONTROVERSIES IN THE USE OF HBO FOR CO POISONING

Hyperbaric oxygenation as a valid treatment for CO poisoning is recognized and supported by several authors in addition to those listed in Table 10-2. These authors include; Dinman (1974), Unsworth (1974), Andersen (1978), Myers *et al.* (1979), Neubauer (1979), Jackson and Menges (1980), Luzhnikov *et al.* (1981), Crocker (1984), Kindwell (1985), Broom and Pearson (1987), Mathieu and Wattel (1985), Jardin (1985), Meredith and Vale (1988).

Several controversies, however, still exist and these will be discussed.

### Normobaric versus hyperbaric oxygen

The use of normobaric oxygen (NBO) in CO poisoning has been described in Chapter 9. The decision whether to use NBO or HBO is also tied in with the decision regarding referral to hyperbaric facilities. Some guidelines to help in this decision are shown in Table 10-3.

TABLE 10-3
HYPERBARIC OXYGEN VERSUS NORMOBARIC OXYGEN FOR CO POISONING
*(From: Colignon and Lamy, 1986. By permission of the authors).*

| | | |
|---|---|---|
| Hyperbaric facilities available | COHb>25% | HBO |
| | COHb<25% | HBO if symptoms, NBO if none |
| No hyperbaric facilities | COHb>40% | Immediate referral to HBO center |
| | COHb<40% no symptoms | NBO |
| | COHb<40% with symptoms | Referral to HBO center |

Hyperbaric oxygen, if available, should be used at COHb levels of 25% or above. The clinical picture of the patient with a history of CO exposure is the deciding factor for the initiation of HBO therapy, and the COHb levels should be a secondary consideration. Due to the cost and limited availability of hyperbaric chambers, a decision regarding transfer of the patient to a hyperbaric facility is not always easy, particularly when the patient is critically ill. If possible, the patient should be transferred to a multiplace chamber with facilities for critical care and suitably qualified personnel. During transport to such a facility, the patient should receive 100% oxygen, suing a mask and taking care that the exhaled air is not rebreathed. The argument that normobaric oxygen is always satisfactory for severe CO poisoning can no longer be sustained. A $pO_2$ of 1800 mm Hg achieved by HBO is definitely more effective than the maximal $pO_2$ of 760 mm Hg attainable by normobaric 100% oxygen. In practice it is much lower than this since few oxygen masks exist that are suitable for administering oxygen to achieve $pO_2$ above 600 mm Hg.

Ducassé *et al.* (1990) carried out a randomized study to compare the effects of normobaric oxygen versus HBO therapy in patients with moderate CO poisoning. In conscious patients without neurological impairment, one HBO treatment at 2.5 ATA for 1 1/2 h, during the first two hours after admission had the following advantages:

1. Faster recuperation from the symptoms such as headache, nausea etc.
2. Quicker elimination of CO during the first two hours. After 12 h, there was no difference in blood COHb levels between the two groups.
3. Less EEG abnormalities after three weeks in the group treated by HBO.
4. Recovery of the cerebral vasomotor response in the HBO treated group, as shown by acetazolamide test.

The authors recommend that HBO should be used in the initial treatment of all patients with CO poisoning regardless of the severity of initial clinical manifestations.

In spite of the proven superiority of HBO over NBO in CO poisoning, several other factors play a part in the process of decision making regarding the transfer of a patient to a hyperbaric facility. Haddad (1986) gave the example of a patient who is conscious and confused, has a COHb level of 25% after NBO treatment but has a blood pH of 7.2 with ECG indicative of myocardial injury. The nearest HBO facility in this case is 75 miles away, accessible only by surface transport and has only a monoplace chamber with no acute care facilities. I would advise against transferring the patient in this situation.

In case of an unconscious apneic victim of CO poisoning who is severely acidotic and hypotensive, with ECG evidence of myocardial ischemia, the decision to transfer the patient to a distant hyperbaric facility should be deferred. Such a patient should be treated in an ICU and given 100% oxygen until the condition is stabilized (Broome 1988).

## Some Critics of HBO Therapy for CO Poisoning

Even those who recognize the value of HBO, question its role in CO poisoning because there are no definite correlations with COHb levels and COHb levels are not a definite guide for therapy. Dolan (1985) admitted that HBO therapy is beneficial in CO poisoning but stated that there are no definite criteria for its use. Olson (1984) presented a critical review of the treatment methods of CO poisoning and questioned if 100% oxygen or HBO can alter mortality or improve neurological improvement. He suggested that carefully controlled prospective studies be done to assess the role of HBO in the treatment of CO poisoning. Another group of skeptics of HBO therapy (Krantz *et al.* 1988) have presented a series of 79 severe cases of CO poisoning managed by conventional methods. The mortality was 30% and 14% of the patients were discharged after a long term hospital treatment with signs of brain damage. Their mortality was far higher than that of the worst series of cases treated by HBO.

In a survey of the use of HBO for CO poisoning in an academic setting, no consensus of opinion was found regarding the merit of HBO as compared to conventional therapy (Roy *et al.* 1989).

A recent study (Raphael *et al.* 1989) carried out a trial of normobaric and hyperbaric oxygen for acute carbon monoxide intoxication in 629 adults who had been poisoned at home in the 12 h preceding admission to the hospital. It was a randomized study with grouping based on whether or not there was initial loss of consciousness. In those without any loss of consciousness HBO was compared with NBO treatments and difference was noticed in the recovery rate. Those who had an episode of loss of consciousness were treated either by a single 2 h session of HBO at 2 ATA followed by 4 h of NBO or by 4 h of NBO with 2 sessions of HBO 6-12 h apart. Two sessions of HBO were shown to have no advantage over single session. The authors concluded that those who do not sustain initial loss of consciousness should be treated by NBO regardless of the COHb levels. The authors did not deny the usefulness of HBO in those with loss of consciousness but stated that 2 sessions of HBO had no advantage over a single session. The methodology in the study is questionable and I have the following criticisms of the study:

1. The follow-up of one month is too short to detect the delayed neuropsychological sequelae of CO poisoning.
2. Too many patients were lost to follow-up (up to 13.7%).
3. There was no group with sham HBO treatment in the patients without loss of consciousness. Obviously such sham treatments would be unethical in the patients with loss of consciousness.
4. There was no neuropsychological assessment which is the most important diagnostic procedure for assessment of sequelae.

It is recommended that future studies should take these shortcomings into account and if possible add neuroimaging methods such as MRI and Spectamine scan for pre- and post-treatment assessment of patients in the study.

### Complications of the Use of HBO

HBO has some complications and these have been described in detail by Jain (1990). Oxygen-induced convulsions are extremely rare when pressures under 3 ATA are used. Barotrauma to the ear can occur and in an unconscious patient it may be advisable to do a myringotomy prior to HBO session. Concern has been expressed that HBO may aggravate the free radical-mediated damage in burns and in ischemic reperfusion injuries. Practical experience has not revealed any evidence of aggravation of such damage by HBO therapy. Safety record of HBO therapy is quite good and undue concern about fires and explosions in hyperbaric chambers as expressed by Kinsella and Clark (1988) is not justified.

Sloan *et al.* (1989) have reviewed their complications in 297 seriously ill cases of CO poisoning treated by HBO. Cardiac arrest occurred in 8% of the patients and in all cases prior to HBO therapy. In 2% of the patients these episodes recurred in the hyperbaric chamber. Other complications included; emesis (6%), seizures (5%), agitations requiring restraint or sedation (2%), tension pneumothorax (1%), tympanic membrane rupture (1%). The authors concluded that none of the complications should deter from the HBO treatments as all of these could be managed in the multiplace chamber.

## TREATMENT OF CO POISONING IN PREGNANCY

In the past pregnancy was considered to be only a relative contraindication for the use of hyperbaric oxygenation, because of the possible toxic effects of oxygen on the fetus. Dangers of hyperoxic exposure to the fetus have been demonstrated in animals. However, these exceeded the time and pressures routinely used in clinical therapy.

Margulies (1986) has shown that 100% oxygen given to pregnant women with CO intoxication should be prolonged to five times of what the mother needs, due to slow elimination of CO by the fetus. Van Hoesen *et al.* (1989) treated CO intoxication (COHb 47.2%) in a 17-year old pregnant female at 37 weeks of gestation using HBO at 2.4 ATA for a 90 min treatment. The patient recovered and produced a healthy baby at full term normal delivery. If the mother had been left untreated, considerable morbidity would have been anticipated for the mother as well as the fetus. They reviewed the literature on this subject and made the following recommendations:

1. Administer HBO therapy if the maternal COHb level is above 20% at any time during the exposure.
2. Administer HBO therapy if the patient has suffered or demonstrated any neurological signs regardless of the COHb level.
3. Administer HBO therapy if signs of fetal distress are present, *i.e.* fetal tachycardia, decreased beat-to-beat variability on the fetal monitor, or late decelerations, consistent with the COHb levels and exposure history.
4. If the patient continues to demonstrate neurological signs or signs of fetal distress 12 h after initial treatment, additional HBO treatments may be indicated.

Careful documentation of experience with this treatment is necessary to determine the long term sequelae and effectiveness of treatment with HBO during pregnancy.

## TREATMENT OF CO POISONING IN CHILDHOOD

CO poisoning in children may be difficult to diagnose (See Chapter 7). Lacy (1981), Binder (1980) and Roberts (1980) have drawn attention to the late neurological sequalae of CO intoxication in children, which are related to hypoxic central nervous system tissue damage. HBO therapy is indicated in CO poisoning in children when symptoms are present and COHb is over 40%. Gozal *et al.* (1985) presented 6 children who were comatose from CO poisoning and all of them recovered without neurological sequelae following HBO therapy.

## TREATMENT OF SMOKE INHALATION INJURY

Smoke inhalation involves multiple toxicities and pulmonary insufficiency, as well as thermal and chemical injury. CO intoxication is the most immediate life-threatening disorder in such cases (Cabalane and Demling 1984). Langford and Armstrong (1989) have proposed an algorithm as a guide to the management of injury from smoke inhalation (Fig. 10-3).

Fein *et al.* (1980) have reviewed the pathophysiology and management of complications resulting from smoke inhalation and recommend 100% oxygen as supplemental treatment. Hart *et al.* (1985) recommend HBO as an adjuvant treatment of smoke inhalation as it is beneficial for both CO and cyanide poisoning resulting from smoke inhalation. Myers *et al.* (1980) suggest the following guidelines for the rescue personnel to direct the patient to an emergency service with a hyperbaric facility.

1. Those who are unconscious
2. Those who are combative but responsive
3. Those not responding to verbal instructions or painful stimuli.

If the patient meets these criteria, 100% oxygen is administered initially during transport to a hyperbaric emergency medical center. If the COHb is over 20% and the surface burns are less than 20%, the patient should be treated initially with HBO and then transferred to a burns center, unless the burns service is located in the hyperbaric facility itself. HBO is given at 2.8 ATA for 46 min using 100% oxygen. Patients with surface burns more extensive than 20% should be treated initially at a burns center. Grube *et al.* (1988) reviewed their experience of treating ten patients with burns and CO poisoning using HBO. They encountered complications such as barotrauma to the ears, metabolic disturbances, ventilatory impairment, and hemodynamic instability leading to cardiac arrest in two patients. They also reviewed the literature on this subject and cast some doubt on the reported incidence of delayed neuropsychiatric sequelae of CO poisoning. They suggested controlled randomized multi-center studies to establish the true inci-

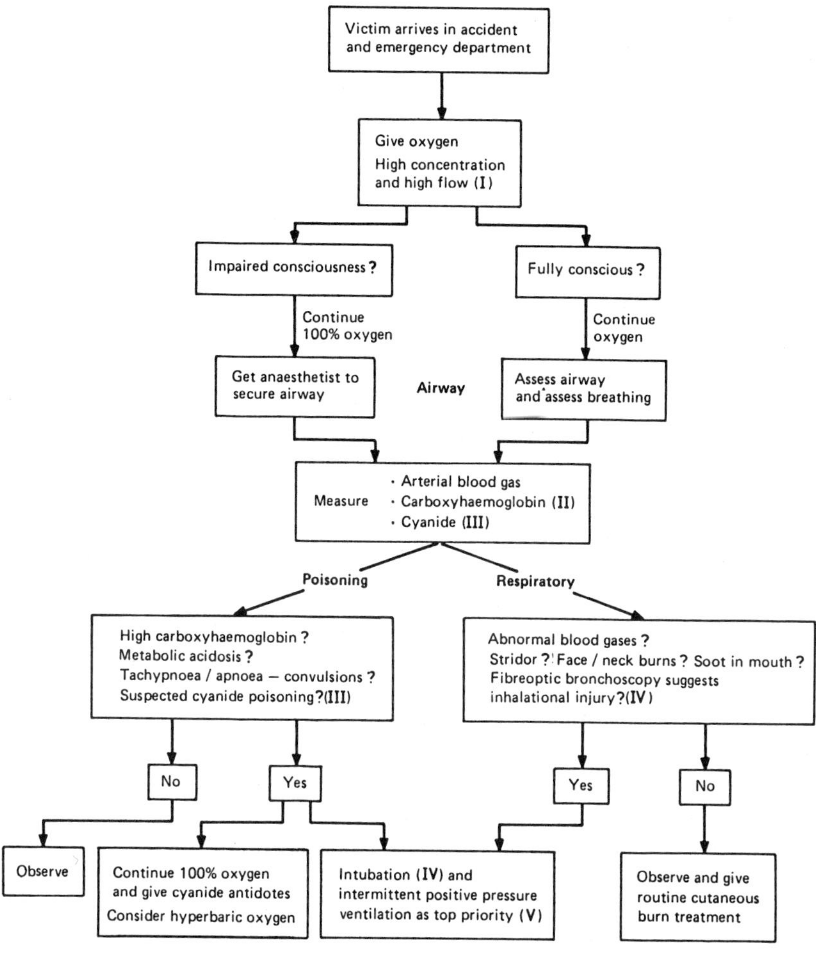


Fig. 10-3: Algorithm for treating the victims of smoke inhalation injury *(Reproduced from: Langford and Armstrong 1989, by permission).*

dence of late sequelae. We consider their patients to be too sick to be put into a hyperbaric chamber with pressure of 3 ATA. Moreover, some of the complications experienced by their patients may not have been due to HBO.

Experimental pulmonary edema due to smoke inhalation is lessened by HBO (Stewart *et al.* 1988). This may be the explanation of benefit of HBO on respiratory insufficiency associated with smoke inhalation and CO poisoning.

Thom (1989) has presented an exhaustive review of the smoke inhalation injury. He has pointed out that CO accounts for more than 50% of the approximately 12000 annual fire-associated deaths. He has recommended that patients who suffer severe CO intoxication should be referred for HBO treatment. HBO therapy has also been recommended for children with CO poisoning associated with smoke inhalation (Parish 1986).

## PREVENTION AND TREATMENT OF LATE SEQUELAE OF CO POISONING

Several reports indicate that the incidence of secondary syndromes is reduced by adequate treatment with HBO in the acute stage of CO poisoning. Myers *et al.* (1985) found no delayed sequelae in any of the 500 patients treated by HBO during the acute stage of CO poisoning. The incidence of these complications was 12% in those treated by normobaric oxygenation.

Takahashi *et al.* (1987) prolonged the HBO treatment in acute CO poisoning until the EEG became normal. This usually took three weeks. They considered that this prevented the late manifestations. Empirical overtreatment has been used in the belief that it would ensure no late sequelae. The half-life of CO bound with cytochrome $a^3$ oxidase, which is the determining factor for late sequelae, is not known. Further research is required to evaluate the CO bound to cytochrome $a^3$ oxidase, so that the necessary duration of HBO treatment can be determined more realistically.

HBO has been used for the treatment of late sequelae of CO poisoning. Funatsu *et al.* (1985) reported a case of apallic syndrome with onset after two symptom-free weeks following the CO poisoning. The patient was treated with 40 daily sessions of HBO and made a complete recovery. EEG abnormalities and CT changes (low density lesions of the globus pallidus) did also resolve.

Kondrashenko (1980) has recommended the use of HBO for treating psychosis due to acute CO poisoning.

## CONCLUDING REMARKS

After reviewing all the evidence and our own experience we conclude that HBO has a definite place in the management of CO poisoning. COHb levels cannot be used as a guide to the treatment. The following are generally accepted as indications for HBO in CO poisoning.

1. Symptomatic patients with COHb over 40%.
2. Pregnant patients in whom COHb levels exceed 20% or when fetal monitoring shows signs of distress.
3. Patients with a history of unconsciousness.

We also use $Mg^{2+}$ which is a physiological calcium antagonist and helps in the prevention of late sequelae of CO poisoning by blocking cellular calcium influx.

## REFERENCES

Adamiec L, Kaminski B, Kwiatkowski H et al (1975) Hyperbaric Oxygen in Treatment of Acute Carbon Monoxide Poisoning.
Anaesth. Resus.Inten.Therap. 3:305-313.

Anderson GK (1978) Treatment of carbon monoxide poisoning with hyperbaric oxygen.
Military Med 143:538-541.

Araki R, Nashimoto I, Takano T (1988) The Effect of Hyperbaric Oxygen on Cerebral Hemoglobin Oxygenation and Dissociation RAte of Carboxyhemoglobin in Anesthetized Rats: Spectroscopic Approach.
Adv Exp.Med. Biol. 222:375-381.

Binder JW (1980) Carbon monoxide intoxication in children
Clin Toxicol 16:287.

Broome JR, Skrine H, Pearson RR (1988) Carbon monoxide poisoning; forgotten but not gone.
Brit J Hosp Med 39:298-300.

Broom JR (1988) How urgent is HBO therapyin CO poisoning (letter).
Brit J Hosp Med 40:234.

Cahalane M, Demling RH (1984) Early respiratory abnormalities from smoke inhalation.
JAMA 251:771-773.

Colignon M, Lamy M (1986) Carbon monoxide poisoning and hyperbaric oxygen therapy. In: Schmutz J (ed) Proceedings of the1st Swiss symposium on hyperbaric medicine.
Foundation for Hyperbaric Medicine, Basel, pp. 41-68.

Crocker PJ (1984) Carbon Monoxide Poisoning, the Clinical Entity and its Treatment: A Review.
Military Medicine 149: 257: 257-9

Dinman BD (1974) The Management of Acute Carbon Monoxide Intoxication.
J Occupational Med 16: 662-4.

Dolan MC (1985) Carbon monoxide poisoning.
Can Med Assoc J 133: 392-9.

Ducassé JL, Izard Ph, Celsis P et al (1990) Moderate Carbon Monoxide Poisoning: Hyperbaric or Normobaric Oxygenation? Human randomized study with tomographic cerebral blood flow measurement.
In: Schmutz J, Bakker D (eds) Proceedings of the 2nd Swiss Symposium on Hyperbaric Medicine, Foundation for Hyperbaric Medicine, Basel, pp. 267-275.

End E, Long CW (1942) HBO in carbon monoxide poisoning. I. Effect on dogs and guinea pigs. J. Ind. Hyg Toxicol 24:302-6.

Fein A, Leff A, Hopewell PC (1980) Pathophysiology and management of the complications resulting from fire and the inhaled products of combustion. Critical Care Medicine 8:94-97.

Funatsu K, Yamada S, Takamuki K et al (1985) Sauerstoff-Überdruckbehandlung der Nachkrankheit bei Kohlenmonoxidvergiftung. Deutsch Med Wschr 110:140-143.

Goulon M, Barois A, Bapin M et al (1969) Intoxication oxycarbonée et anoxie agiue par inhalation de gaz de charbon et d'hydrocarbures. Ann Méd interne 120: 335-349.

Gozal D, Ziser A, Shupak A et al (1985) Accidental Carbon Monoxide Poisoning. Clinical Pediatrics 24: 132-135.

Grube BJ, Marvin JA, Heimbach DM: Therapeutic Hyperbaric Oxygen: Help or Hindrance in Burn Patients with Carbon Monoxide Poisoning? JBCR 9: 249-252.

Hadda LM (1986) Carbon monoxide poisoning: to transfer or not to transfer? Ann Emerg Med 15:1375.

Hart GB, Strauss MB, Lennon PA et al (1986) Treatment of smoke inhalation by hyperbaric oxygenation. J Emerg Med 3:211-215.

Haldane JS (1895) The relation of action of carbonic oxide to oxygen tension. J Physiol (Lond) 18:201-217.

Hyperbaric Center Advisory Committee, EMS, City of New York (1988) A registry for carbon monoxide poisoning. Clin toxicol 26:419-441.

Jackson DL, Menges H (1980) Accidental carbon monoxide poisoning. JAMA 243:772-774.

Jain KK (1990) Textbook of hyperbaric medicine. Hogrefe & Huber, Toronto-Bern.

Jardin F (1985) L'oxygénothérapie hyperbare au cours des intoxications oxycarbonées. La Presse Médicale 14:283.

Kienlen J, Alardo JP, Dimeglio G et al (1974) Traitement de l'intoxication oxycarbonée par l'oxygène hyperbare - à propos de 370 observations. J de Mèdecine de Montpellier 9:237-243.

Kindwall EP, Goldman RW (1988) Hyperbaric Medicine Procedures. St. Luke's Hospital, Milwaukee.

Kindwall EP (1985) Hyperbaric Treatment of Carbon Monoxide Poisoning. Annals of Emergency Medicine 14: 1233-1234.

Kindwall EP (1980) Carbon monoxide poisoning and cyanide poisoning. Hyperbaric Oxygen Rev 1:115-122.

Kinsella J, Clark CJ (1988) Hazards of HBO therapy. Brit J Hosp Med 40:406-407.

Krantz T, Thisted B, Strøm J et al (1988) Acute carbon monoxide poisoning. Acta Anesth Scand 32:278-282.

Kondrashenko VT (1980) Hypoxia in acute exogenous psychoses and its treatment. Zh Nevropatol Psikhiatr 80: 898-904.

Koyama K (1976) Acute experimental carbon monoxide poisoning and hyperbaric oxygenation (HB=) Tokyo Jikeikai Med J 91: 195-215.

Lackey DJ (1981) Neurologic sequelae of acute carbon monoxide intoxication. Am J Dis Child 135:145.

Langford RM, Armstrong RF (1989) Algorithm for managing injury from smoke inhalation. BMJ 299:902-905.

Luzhnikov EA, Isakov Yuv, Lukin-Butenko GA (1981) Hyperbaric oxygenation in acute carbon monoxide poisoning. Klin Med (Moscow) 59:89-93.

Margulies J (1986) Acute carbon monoxide poisoning during pregnancy. Am J Emerg Med 4: 516-19.

Mathieu D, Nolf M, Durocher A, et al. (1985) Acute Carbon Monoxide Poisoning Risk of Late Sequelae and Treatment by Hyperbaric Oxygen. Clinical Toxicology 23:315-24.

Mathieu D, Wattel F (1985) A propos du traitment par oxygénothérapie hyperbare au cours des intoxications oxycarbonées. La Presse Medicale 14:1433.

Meredith T, Vale A (1988): Carbon monoxide poisoning. BMJ 296:77-79.

Myers RAM, Synder SK, Emhoff TA et al (1985) Subacute Sequelae of Carbon Monoxide Poisoning. Annals of Emergency Med 14: 1163-67.

Myers RAM (1980) Treatment protocol for smoke inhalation victims. Occupational Health and Safety 49: 56-60.

Myers RAM, Linberg SE, Cowley RA et al (1979) Carbon monoxide poisoning; the injury and its treatment. J Am Coll Emerg Physicians 8:479-484.

Neubauer RA (1979) Carbon monoxide and hyperbaric oxygen. JAMA 139:629.

Norkool DM, Kirkpatrick JN (1985) Treatment of Acute Carbon Monoxide Poisoning with Hyperbaric Oxygen: A Review of 115 Cases.
Annals of Emerg Med 14: 1168-1171.

Olson KR (1984) Carbon monoxide poisoning; mechanisms, presentation, and controversies in management.
J Emerg Med 1:233-243.

Pace N, Strajman E, Walker EL (1950) Acceleration of carbon monoxide elimation in man by high pressure oxygen.
Science 111: 652-4.

Parish RA (1986) Smoke inhalation and carbon monoxide poisoning in children.
Ped Emerg Care 2:36-39.

Radushkevich VP, Koroteeva EI (1976) Parallel blood circulation with oxygenation of blood in severe poisoning with carbon monoxide fumes (Russian).
Vestn Khir 116:131-134.

Raphael JC, Elkharrat D, Jars-Guincestre MC et al (1989) Trial of normobaric and hyperbaric oxygen for acute carbon monoxide intoxication.
Lancet ii:414-419.

Rioux JP, Myers RAM (1989) Hyperbaric oxygen for methylene chloride poisoning.
Ann Emerg Med 18:691-695.

Roy TM, Mendieta JM, Ossorio MA et al (1989) Perceptions and utilization of hyperbaric oxygen therapy for carbon monoxide poisoning in academic setting.
J Kentucky Med Assoc 87:223-226.

Sasaki T (1975) On Half-Clearance Time of Carbon Monoxide Hemoglobin in Blood During Hyperbaric Oxygen Therapy (OHP)
Bull Tokyo Dent Univ 22:63-77.

Sharp GR, Ledingham IM, Norman JN (1962) The application of oxygen at 2 atmospheres pressure in the treatment of acute anoxia.
Anaesthesia 17:136-144.

Sluijter ME (1963) The treatment of carbon monoxide poisoning by the administration of oxygen at high pressure. Springfield, Illinois, Charles C. Thomas.

Sluitjer ME (1967) The treatment of carbon monoxide poisoning by administ/ration of oxygen at high pressure. In, Bour H and Ledingham IMC (eds) Carbon Monoxide poisoning, Elseview, Amsterdam/London/New York. pp 123-182.

Smith G, Ledingham IM, Sharp GR et al (1962) Treatment of Coal-Gas Poisoning with Oxygen at 2 Atmospheres Pressure.
Lancet 1: 816-818.

Sloan EP, Murphy DG, Hart R et al (1989) Complications and protocol considerations in carbon monoxide-poisoned patients who require hyperbaric oxygen therapy.
Ann Emerg Med 18:629-634.

Stewart RJ, Yamaguchi KT, Samadani S et al (1988) Effects of oxygen and pressure on extravascular lung water following smoke inhalation. J Hyperbaric Med 3:173-178.

Sukoff MH, Hollin SA, Espinosa OE et al (1968) The protective effect of hyperbaric oxygenation in experimental cerebral edema. J Neurosurg 29:236-241.

Takahashi H, Kobayashi S, Sakakibara K (1987) The value of hyperbaric oxygen therapy in preventing late manifestations following acute carbon monoxide poisoning. In: Kindwall EP (ed) Proc. Eighth Int Cong. Hyperbaric Medicine, San Pedro, California, Best Publishing Company pp. 274-281.

Thom SR (1989) Smoke inhalation. Emergency Med Clinics of North America 7:371-387.

Timchuk LD, Luchko AS, Palamarchok VN et al (1981) Hyperbaric oxygen in complex treatment of acute poisoning. In: Yefuny SN (ed) Proceedings of the 7th international congress of hyperbaric medicine. USSR Academy of Sciences, Moscow pp, 333-334.

Tirpitz D, Baykara T (1988) Hyperbare Oxygenation bei CO-Intoxication. Der Informierte Arzt 8:51-54.

Unsworth IP (1947) Acute CArbon Monoxide Poisoning. Anaesth Inten Care 4:329-332.

Valois JD, Schàde JP (1967) an electrophysiological study of histotoxic anoxia under normal and hyperbaric conditions. In, Bour H and Ledingham IMC: Carbon Monoxide Poisoning, Elsevier, Amsterdam/London/New York. pp 183-197.

Van Hoesen KB, Camporesi EM, Moon RE (1989) Should hyperbaric oxygen be used to treat the pregnant patient for acute carbon monoxide poisoning? JAMA 261:1039-1043.

Welsh F, Matos L, DeTreville RTP (1980) Medical Hyperbaric Oxygen therapy. Aviation Space Environment Med 51:611-614.

Winter A, Shatin L (1970) Hyperbaric oxygen in reversing carbon monoxide coma. New York State J Med 70:880-884.

Yee LM, Brandon GK (1983) Successful Reversal of Presumed Carbon Monoxide-Induced Semicoma. Aviation, Space and Environmental Med 54: 641-3.

Youn BA, Kozikowski RJ, Myers RAM (1989) The development of treatment algorithm in methylene chloride poisoning based on multicase experience (abstract). Undersea Biomed Res 16(suppl):20

Yun DR, Cho SH (1983) Hyperbaric oxygen treatment in acute CO poisoning. Korean J Preventive Med 16:153-156.

Ziser A, Shupak A, Halpern P et al (1984) Delayed hyperbaric oxygen treatment for acute carbon monoxide poisoning. Brit Med J 289:960.

# Chapter 11

## PREVENTION OF CO POISONING

### INTRODUCTION

Prevention is the most important aspect of dealing with accidental carbon monoxide poisoning. Prevention is a public health as well as an individual's responsibility. Preventive measures depend on the socioeconomic and politcal conditions. Public eduation is important in order to reduce the incidence of accidental carbon monoxide poisoning. The recommendations can be stated briefly as follows.

### PRECAUTIONS AT HOME

1. All home appliances that burn fuel should be properly installed and operated with routine maintenance and inspection at regular intervals.
2. Fireplaces and wood burning stoves should be adequately ventilated.
3. Gas cookers should not be used for space heating.
4. Charcoal should not be burned indoors or in other enclosed spaces.

### PRECAUTIONS IN AUTOMOBILE AND ROAD TRAFFIC

1. The motor vehicle exhaust system should be checked regularly for any leaks. Catalytic converters should be installed to reduce CO emission. CO levels should be checked in the passenger compartment with windows closed to see whether they exceed safety levels.
2. Traffic jams should be avoided particularly in tunnels and if one hits them accidentally, the car engine should be stopped and car windows closed. Once the obstruction is removed, cars should start one after the other. The ventilation system in the tunnels should be improved.
3. Persons exposed to heavy traffic, such as traffic policemen should undergo periodical psychometry and COHb estimations. If an impairment of mental function is detected or there is an increase of COHb over 5%, the persons concerned should be temporarily taken off street duty until the COHb levels have returned to below 5%.
4. No jogging or bicycling on streets with heavy automobile traffic.

## PRECAUTIONS AT THE WORK PLACE

1. Factories where there is risk of exposure to CO should apply and monitor safety precautions rigorously.

2. Mines are particularly dangerous. In coal mines, CO levels must be monitored periodically and the miners screened by using COHb measurements. Fatal and non-fatal coal mining injuries have declined in the USA, not only due to state regulatory and enforcement activity but also due to collective capacities of miners to organize effectively against unsafe conditions (Wallace 1987).

3. Change of process to reduce CO formation. The following are some examples.

   i. Battery operation for towing of aircraft and forklift trucks that are operated in confined spaces.
   ii. Electric traction for urban and underground railway system.
   iii. Multiple-chamber refuse burning to replace the open burning of refuse.
   iv. Electrically powered conveyer belts to move cars through inspection lanes. Similar devices are currently in use at car washes.

4. Control of CO formation at source in industrial processes. This involves an appropriate manipulation of the four main variables influencing CO production; fuel/air ratio, flame temperature, gas residence time, and turbulence in the combustion chamber (Shepherd 1983). An increased air supply can reduce the output of carbon monoxide from a foundry cupola. Improved methods of producting iron and steel from pig iron can reduce CO emissions from 1000 kg per ton to less than 10 kg per ton.

5. Ventilation. If a particular area of the work place is known to be heavily contaminated, appropriate ventilation must be arranged. Natural ventilation, with its many variables, is not adequate for this purpose. Dilution ventilation, such as by forcing air into or from a room by a fan is more effective. It is important to ensure that the intake air is reasonably free from CO.

Local ventilation is the appropriate method for controlling the high concentrations of CO found in some industrial workplaces.

6. Protective equipment. In some occupations such as mining and firefighting, potential exposure to CO is so great that protective equipment must be provided. Large quantities of CO cannot be removed satisfactorily by a filter. Rescue workers are sometimes provided with a closed-circuit type of breathing equipment. Oxygen can be added to this system from a portable cylinder.

The current safety standards for exposure of workers to CO are listed in Chapter 3. The safety margin of these is quite slender and some experts insist on more exacting limits. These standards offer insufficient protection against the acceleration of CO absorption by physical activity and transient peaks of pollution.

## CONTROL OF ENVIRONMENTAL POLLUTION IN THE CITIES

Since the main source of urban CO is the automobile, the key to control is to reduce the number of gasoline-driven vehicles entering large cities. The entry of cars into a city can be restricted by several measures, some examples of which are:

i. Requirement of costly permits to enter a city.
ii. Restriction of parking places in the city.
iii. Providing parking places outside the city and public transport from there to the city center.
iv. To designate the city core as traffic free areas.
v. To restrict entry to city to vehicles carrying four or more passengers.

The usual practice of assessing atmospheric pollution in a city by sampling CO concentrations of the outdoor air at a few spots is of questionable value in predicting acutal exposures to CO (Wallace and Ziegenfus 1985).

## SMOKING CESSATION

Cigarette smoking is the single most common and most important cause of chronic low grade CO poisoning (see Chapter 4). The only preventive measure is to give up smoking. Simple reduction of the number of cigarettes smoked is not enough. Benowitz *et al.* (1986) found that by reducing from 37 to 5 cigarettes per day, the intake of tobacco toxin increased roughly three-fold and the daily exposure to CO declined only by 50%. Smokers regulate their intake of nicotine by adjusting the pattern of inhaling.

The danger of chronic low grade poisoning in smokers is that, when they are exposed to further acute atmospheric pollution with CO, such as in heavy road traffic, they are liable to develop high COHb levels, which may manifest themselves clinically. Non-smokers exposed to similar pollution environments may not show toxic effects. Pregnant women who smoke are at particular risk of CO poisoning and should be strongly advised to give up smoking. Smoking cessation has been shown to lessen the risk of death or myocardial infarction in older as well as younger persons with coronary artery disease (Hermanson *et al.* 1985). Impact of cigarettes on health is considerable as 28.8% of adults smoke (MMWR 1989).

Passive smokers are exposed to the exhaled smoke of smokers in a closed room. Fresh air inflow recommended by Wanner and Schatter (1987) is 12=15m$^3$ per person per hour for CO and 30-70 m$^3$ per person per hour for smoke. Because urban dwellers spend 90% of their time indoors, efforts to control effects of pollutants such as CO should be directed to indoor as well as atmospheric pollution (Spengler and Sexton 1983).

## Nicotine addiction

Most of the harmful effects of cigarette smoking are due to the tar and CO content of tobacco smoke. The addiction, however, is due to nicotine. Biology of nicotine dependence was the topic of Ciba symposium held in London, England, in November, 1989. Approaches to smoking cessation are based on the treatment of nicotine addiction. Nicotine is a ganglionic stimulant although it has a depressant effect as well. Many of the effects of nicotine are mediated through increased release of ACh and noradrenaline. The cardiovascular effects of nicotine include increase in blood pressure, heart rate, myocardial oxygen consumption, coronary blood flow and peripheral vasoconstriction. Cerebral effects include increase of cerebral blood flow and CMRGI. There is behavioral arousal and mental performance is stimulated. Chronic nicotine use leads to increase in the number of nicotine receptors in the brain.

Nicotine addiction is similar to addiction to alcohol and heroin. Symptoms of withdrawal include; craving, irritability, frustration, anger, anxiety, poor concentration, restlessness, weight gain, and decreased heart rate. Craving is the most characteristic feature of withdrawal and reaches a peak within 24 hours after nicotine cessation and may continue for several months. Craving occurs in about 47% of the smokers after withdrawal.

## General Measures for Smoking Cessation

Simple techniques of behavioral intervention and counseling may produce success rates of 5% in smoking cessation. The approach to the problem of nicotine addiction is multidisciplinary and an overview of the measures is shown in Table 11-1. General pharmacotherapy for smoking cessation has been reviewed by Nunn-Thompson and Simon (1989). None of the methods used in the past have been entirely satisfactory and most of the chronic heavy smokers return to their habit. Most success has been achieved so far with techniques based on some form of nicotine replacement.

The usual techniques of nicotine replacement therapy are: nicotine chewing gum, nicotine nasal spray, and transdermal nicotine. Pooled data from four nicotine chewing gum trials (Lam *et al.* 1987) showed a success rate of 35% when combined with psychotherapy in a three-month course of treatment. The major disadvantages of the chewing gum preparation are high peaks of nicotine in plasma and local irritation of the buccal mucosa.

For planning replacement therapy, it should be noted that one cigarette delivers about 1.2 mg nicotine and the effect is felt in about 7 s. Nicotine from the cigarette is delivered in about 10 puffs. If nicotine is given intravenously as 1.78 mg of it dissolved in 10 ml of normal saline, it takes 14 s to achieve comparable concentration of nicotine with smoking a cigarette. If one cigarette is smoked per hour and profile of blood nicotine is studied, steady state is reached in a few hours. Smokers regulate fluctuations in nicotine levles up to two-third of the peaks. If the

TABLE 11-1
CLASSIFICATION OF MEASURES FOR SMOKING CESSATION

A. Psychotherapy
Individual counselling
Group therapy
Relaxation training
Hypnosis

B. Pharmacological
Nicotine replacement
-nicotine spray
-nicotine chewing gum
-transdermal nicotine systems
-inhalation of nicotine impregnated cigarettes (unlighted)
-low resistance filter cigarettes

Symptomatic treatment of withdrawal: clonidine and tricyclic anti-depressants
Nicotine antagonists: oral mecymyalmine
Smoking deterrants: silver nitrate

C. Miscellaneous
Acupuncture
Aversion training

level falls below one-third of the peak, craving is experienced. A steady level of nicotine is thus required to prevent craving.

## Transdermal Nicotine

To date the most effective method of nicotine replacement therapy is the Transdermal nicotine patch (Nicotell TTS) developed by CIBA-GEIGY, Basel and approved in Switzerland. It is available in sizes of 10, 20, and 30 $cm^2$ delivering 7,14 and 21 mg of nicotine respectively per 24 hours. The patch is replaced once a day and maintains a more constant level of nicotine in the blood than the chewing gum thus reducing craving. Some patients develop cutaneous reactions to the patch, but these ususally subside after removal of the patch.

## Pharmacology and Pharmacodynamics of Transdermal Nicotine

Bannon *et al.* (1989) studied the transdermal delivery of nicotine in normal human volunteers following 15-60 mg dosage range. The levels achieved were in the same range as reported for smokers and were sustained for 24 hours. No significant accumulation of nicotine occurred following multiple dosage application of 30-mg patches.

Dubois *et al.* (1989) have investigated the bioavailability of nicotine in healthy human volunteers following single and repeated administration of different doses of TNS. They determined plasma and urine nicotine as well as cotinine levels by high performance liquid chromatography. Their findings were:

1. The 24-h plasma nicotine concentration reached after TNS application compared well with the plasma nicotine footpoints- not the peaks- observed in moderate to heavy smokers.

2. One or two hours after the removal of the systems, there was a very slow decline of the nicotine concentrations. After repeated applications of TNS, there was evidence for only a limited nicotine rise in plasma and urine over 10 days. The steady state of nicotine was reached in four days.

3. Continuous delivery of nicotine over 24 hours resulted in an early morning plasma concentration, which would prevent the craving for the first cigarette in case of a smoker.

Müller *et al.* (1989) investigated the pharmacology of transdermal nicotine in relation to cigarette smoking and their conclusions were:

1. Cardiovascular changes induced by smoking were most pronounced after the first cigarette following 10 hour's abstinence. Compared with these, the changes induced by TNS were less pronounced. A slight rise of blood pressure was detectable only on the first day of application and disappeared after 10 days application.

2. In contrast to cigarette smoking, TNS had no effect on the cutaneous blood flow and skin temperature.

A disparity between the effects of nicotine and cigarette smoking can be explained by the presence in the cigarette smoke of substances such as CO which may modify the effects of nicotine.

Piraino *et al.* (1989) tested the TNS delivery system in ten smokers. No smoking was allowed during the study. Delivery reate from the 10 cm nicotine patch was 0.8 mg/h. A steady state of blood nicotine level comparable to the pre-treatment level was achieved by day 5. An increase in the mean resting heart rate of 10 beats/min was observed during the course of the study. Nicotine was well tolerated by the subjects.

## Clinical Trials

Clinical studies of the use of transdermal nicotine delivery for smoking are shown in Table 11-2. The studies by Abelin *et al.* (1989) involved the use of the Nicotell TTS developed by CIBA-GEIGY, Basel, but no psychotherapy. The better results in the 1989b study are explained by the better motivation of the university students who were younger and also better motivated. A third study is under way by Prof. Abelin of Bern (personal communication, November 1989) which will test the effects of transdermal nicotine on smoking cessation in Swiss physicians. The patches in these studies are well tolerated except that 25% of the patients in the nicotine group and 13% of the patients in the placebo group had to discontinue the patches due to skin reactions. The patch which is usually applied to the skin of the lower abdomen usually holds well in place for 24 hours if applied on a dry skin. Daily activities such as exercise and showers usually do not dislodge the patch.

TABLE 11-2

| *Author and year*<br>*Country* | *Number of subjects* | *Method of study* | *Results and comments* |
|---|---|---|---|
| Buchkremer et al. (1988 a,b)<br>West Germany | 131 | Double-blind prospective study. Random assignment to 3 groups each of which underwent 9 w of smoking cessation plus 6 w of:<br><br>1) nicotine patches daily or<br>2) placebo patches or<br>3) behavioral training | 3, 6, and 12 month follow-up. Significantly higher abstinence rates in the transdermal group than in the other two groups. Therewere no serious side effects. |
| Abelin et al (1989 a) | 199 | 3 m randomized, placebo-controlled. Patients pooled from several general practices.<br>100 patients…nicotine group<br>99 patients…placebo patches | *Abstinence achieved*<br>Treated group Placebo group<br>1 m 41% 19%<br>2 m 36% 20%<br>3 m 36% 23%<br>6 m 22% 12.2% |
| Abelin et al (1989 b) | 112 | Similar protocol as above.<br>56…nicotine group<br>56…placebo group<br>University-based study. Younger subjects (students) who volunteered for the study | *Abstinence achieved*<br>Treated group Placebo group<br>3 w 51.8% 28.6%<br>6 w 42.9% 25%<br>9 w 39.3% 19.6% |

The transdermal nicotine systems is currently expected to be approved soon for general use on a physician's prescription. It will provide a safe and convenient method for treating the nicotine withdrawal and it makes it easier for the patient to get over psychosocial hurdles and improves the response to behavioral therapy. Transdermal nicotine should be integrated into a 4-step package for smokers:

1. Bring the topic to the attention of the smoker.
2. Prepare the act of stopping.
   a) Transdermal nicotine
   b) Formulation of objectives and techniques
4. Follow-up for at least one year.
   a) Special support after starting use of the nicotine patch
   b) Support of continuous abstinence
   c) Support in situation of relapse.

## CONCLUDING REMARKS

With the best efforts, it may not be possible to achieve a 100% success rate in smoking cessation. Some smokers may not want to quit or participate in smoking cessation programs. For those who can achieve abstinence with continuous nictotine substitution therapy, I would consider investigating long term therapy with transdermal nicotine. The major health hazards of smoking, those due to tar and CO, are eliminated by this and an individual has only to bear the effects of nicotine. It is possible that the effects of nicotine administered by transdermal method may be less harmful than that by other methods which involve peaks of blood levels. This subject needs further investigation.

Nictoine is a drug and the medical profession and the drug industry is responsible for its safe use. Cigarette smoke is far more harmful, but cigarett sale is not subject to permission from health authorities. With continued pure nicotine use, the individual is not polluting the surrounding and endangering the health of others. However, the tobacco industry, who claim to have a monopoly on the source of nicotine, do not want any competition in the free sale of its substitutes. These are some of the problems that may have to be resolved by the combined effort of the medical profession, the health authorities, the politicians and the legislators.

## REFERENCES

Abelin T, Müller P, Buehler A et al (1989a) Controlled trial of transdermal nicotine patch in tobacco withdrawal.
Lancet i:7-10.

Abelin T, Ehrsam R, Bühler-Reichert A et al. (1989b) Effectiveness of a transdermal nicotine system in smoking cessation studies.
Meth and Find Exp Clin Pharmacol 11:205-214.

Bannon YB, Corish J, Devane JG et al (1989) Transdermal delivery of nicotine in normal human volunteers.
Europe J Clin Pharmacol 37:285-290.

Benowitz NL, Jacob P, Koslowski LT et al (1986) Influence of smoking fewer cigarettes on exposure to tar, nicotine, and carbon monoxide.
New Engl J Med 315:1310-1313.

Buchkremer G, Bents H, Minneker E et al. (1988a) Langfristige Effekte einer Kombination transdermaler Nikotinzufuhr mit Verhaltenstherapie zur Rauchentwöhnung.
Nervenarzt 59:488-490.

Buchkremer G, Bents H, Minnekar E et al (1988b) Combination of behavioral smoking cessation therapy with transdermal nicotine substitution: long term effects In: M. Aoki, Hisamachi S, Tominaga S (eds) Smoking and Health 1987.
Amsterdam-New York, Excerpta Medica pp. 857-860.

Ciba Foundation Symposium: The biology of nicotine dependence.
Symposium #152, held in London, November 6-9, 1989, To be published.

Dubois JP, Sioufi A, Müller P et al (1989) Pharmacokinetics and bioavailability of nicotine in healthy volunteers following single and repeated administration of different doses of transdermal nicotine systems.
Meth and Find Exp Clin Pharmacol 11:187-195.

Hermanson B, Omenn GS, Kronmal RA et al (1988) Beneficial six-year outcome of smoking cessation in older men and women with coronary artery disease.
New Engl. J Med 319:1365-1369.

Lamm W, Sze PC, Sacks HS et al (1987) Metaanalysis of randomized controlled trials of nicotine chewing gum.
Lancet ii:27-30.

MMWR: Tobacco use by adults —United States, 1987. 38:685-687.

Müller P, Imhof PR, Mauli D et al (1989) Human pharmacological investigations of a transdermal nicotine system.
Meth and Find Exp Clin Pharmacol 11:197-204.

Nunn-Thompson CL, Simon PA (1989) Pharmacotherapy of smoking cessation.

Clinical Pharmacy 8:710-720.

Piraino A, Morgan J, Saris S et al (1989) Pharmacokinetics of multiple dose transdermal nicotine. J Clin Pharmacol 29:9(abstract)

Spengler JD, Sexton K (1983) Indoor air pollution: a public health perspective. Science 221:9-17.

Shepherd RJ (1983) Carbon Monoxide —the silent killer. Springfield, Thomas.

Wallace M (1987) Dying for coal: the struggle for health and safety conditions in American coal mining 1930-1982. Social Forces 66:336-365.

Wallace LA, Ziegenfus (1985) Comparison of carboxyhemoglobin concentrations in adult non-smokers with ambient carbon monoxide levels. J Air Pollution Control Association 35:944-949.

Wanner HU, Schlatter J (1987) Raumluftverunreinigungen- Gesundheitliche Gefährdungen und Massnahmen. Sozial- und Präventivemedizin 32:249-250.

# Chapter 12

## CONCLUSIONS AND DIRECTIONS FOR FUTURE RESEARCH

Review of the current state of the art of CO management leads to the following conclusions:

1. The danger to health of carbon monoxide is very much under-estimated. Chronic occult CO poisoning is far more prevalent than indicated by the published figures which refer mainly to acute cases of poisoning.

2. Smoking is the most common cause of chronic low intensity CO intoxication. All chronic smokers should undergo COHb and neuropsychological testing. If COHb levels are above 5% and there is impairment of mental function, smoking cessation should be strongly advised. Chronic smokers are at a greater risk of fatal CO poisoning when exposed acutely to moderate levels of CO. Smoking cessation is the most important preventive measure against chronic CO poisoning for the individual

3. The automobile is the most common source of CO poisoning outside the home, both acute as well as chronic. Gas heating appliances are the most common source of CO poisoning in the home. Appropriate preventive measures should be directed at these sources.

4. Hypoxia plays an important role in the pathophysiology of CO poisoning. The controversy whether COHb formation is entirely responsible for the effects and the role of cellular respiratory enzymes is not settled as yet.

5. Chronic CO poisoning may mimic several other diseases. A careful review of the history and awareness of CO poisoning as a cause increases the chances of detecting it. This should be confirmed by appropriate diagnostic procedures and COHb determinations.

6. COHb levels do not always correlate with the clinical manifestations.

7. Hyperbaric oxygen therapy is the best treatment for certain cases of CO poisoning. In patients with an episode of loss of consciousness and those with COHb levels of 40% or over and clinical manifestations of CO poisoning, there is no controversy regarding the use of HBO. HBO therapy may be applied even if COHb values are not available but there is a clinical suspicion supported by the history of exposure to CO. There is evidence that prompt and adequate treatment with HBO in the acute stage prevents the late sequelae of CO poisoning. In a symptomatic patient, I would consider a COHb level of 20% significant in recommending HBO treatment.

## Directions for future research

Even though there is considerable information available on this subject and an effective therapy in the form of HBO, further research is required. I consider the following aspects to be worthy of further investigations.

1. Methods for measurement of CO at cellular cytochrome level and measurement of the half life of CO bound there. The information obtained may clarify the mechanisms of CO toxicity and also serve as a guide to the duration of HBO therapy.

2. Refinements in diagnostic procedures. Brief but reliable psychological tests need to be devised for the bedside examination of patients with chronic subclinical CO poisoning. The conventional psychological tests are not practical in these situations.

Metabolic studies of the living brain may reveal useful information in cases of CO-encephalopathy and may serve as a guide to therapy. Spectamine scan may be a useful procedure in such cases.

3. Further research needs to be done for smoking cessation as a preventive measure for chronic CO poisoning. Long term use of nicotine replacement by transdermal nicotine needs to be explored.

4. Further epidemiological studies of CO levels in the atmospheres of urban areas need to be continued particularly to assess the effects of the improved measures against CO-pollution such as widespread use of catalytic converters in the automobiles.

5. Interaction of chronic low grade CO poisoning with therapeutic drugs has not been investigated. There are few studies available about the interaction of smoking with other drugs. Drugs which depend upon the cellular enzyme systems such as P-450 and cytochrome oxidae may undergo altered metabolism in presence of CO bound to these enzymes.

6. Testing of the theoretical models of elimination of CO under HBO conditions. Animal experimental studies as well studies in human patients are required to test such models as the one proposed by Britten and Myers (1985).

## REFERENCE

Britten JS, Myers RAM (1985) Effects of hyperbaric treatment on carbon monoxide elimination in humans.
Undersea Biomed Res 12:431-438.

# INDEX


## A

Aggravating factors for CO poisoning, 56
- aging, 57
- anemia, 57
- high altitude, 57
- pregnancy, 58
- other toxins, 56

Anoxemia in CO poisoning
Anemia in CO poisoning
Atmospheric CO 44, 47
- automobile as source, 44, 45
- chemical reactions, 48
- cigarette smoking, 69
- concentrations, 46
- cycle, 50
- natural sources, 46
- removal, 48
- sinks, 48, 49

Automobile as source of CO, 44

## B

Biochemistry of CO poisoning, 2
Blood gases in CO poisoning, 131
Body stores of CO, 22

## C

Carbon monoxide
- binding to hemoglobin, 28
- biochemistry, 2
- body stores, 22
- cellular effects, 34
- elimination, 30
- endogenous production, 23
- environmental pollution, 44
- epidemiology, 54
- interactions with other toxins, 36
- medical uses, 36
- physical properties, 20
- site of action, 35
- uptake and distribution, 24

Carbonic oxide (see Carbon monoxide)
Carboxyhemoglobin, 28
- half life, 31

Causes of CO poisoning, 55
Chemical reactions of CO, 48
CO (see carbon monoxide)
CO-diffusion capacity, 37
CO-induced hypoxia, 32
CO-poisoning
- causes, 55

chronic CO poisoning, 114
- classification of severity, 114
- clinical features, 113
- clinical diagnosis, 115
- clinicopathological correlation, 122
- delayed effects, 120, 121
- diagnosis, 128-137
- neuropsychology, 117-119
- neurological effects, 119
- occult, 116

Cytochrome oxidase, 33
- binding to CO 35

## D

Diagnosis of CO poisoning
- children, 116-117
- clinical, 113
- laboratory, 129
- pitfalls, 115

DPG (2,3, diphosphoglycerate)
- in chronic hypoxia, 33

Drugs
- adjunct therapy for CO-poisoning, 139
- for treatment of smoking, 168
- interacting with CO 59

## E

Effects of CO on the human body, 84
- cardiovascular, 85-89
- exercise physiology, 93
- gastrointestinal tract, 94
- hematological effects, 90
- kidney, 94
- liver, 94
- nervous system 91-92
- reproductive system, 95-97
- respiratory system, 92-93
- skin, 95

Environmental CO
- safety standards, 51
- measuring methods, 51
- indoor concentrations, 52

Epidemiology of CO poisoning, 54

## H

Haldane
- carbonic oxide paper, 6
- constant, 28
- experiments with hypoxia, 12
- HBO treatment of CO poisoning, 14
- law, 28

HBO (see hyperbaric oxygen)

Hemoglobin
- binding to CO, 28

History of carbon monoxide
- classical works, 5
- historical benchmarks, 3

Hyperbaric oxygen therapy for CO-poisoning, 147-159
- children, 156
- clinical reports, 150-151
- complications, 154
- controversies, 154
- critics, 153
- HBO versus normobaric oxygen, 152
- experimental evidence, 148
- pregnancy, 155
- prevention of late sequelae, 158
- rational, 147
- smoke inhalation, 156-157
- technique, 149

## L

laboratory tests for CO-poisoning
- arterial blood gases, 131
- cardiopulmonary investigations, 132
- CT scan, 133
- COHb, 129
- exhaled CO, 130
- neuropsychological tests, 132
- postmortem COHb, 130

## M

Management of CO poisoning
- chronic occult form, 144
- guidelines, 138
- hyperbaric oxygen, 147-159
- miscellaneous methods, 113
- of associated injuries, 143-144
- of sequelae, 142
- overview, 139
- oxygen therapy, 140
- past therapies, 140
- resuscitation in acute phase, 138
- triage decision chart, 141

Medical uses of CO, 36

Myoglobin, 28

## N

Natural sources of CO, 46

Nicotine
- addiction, 167
- effects, 65
- transdermal, 168-170

Nicotell TTS, 168

Neuropsychology of CO poisoning, 117-119

## O

Oxygen
- dissociation curve in CO-poisoning, 30
- partial pressure in CO-poisoning, 29
- transport and metabolism in CO-poisoning, 31

## P

Pathology of CO poisoning
- fetus, 110
- heart, 109
- kidney, 109
- liver, 109
- lungs, 109
- muscles, 109
- neuropathology, 104-108
- peripheral nerves, 109
- reproductive system, 95-97
- skin, 110

Pathophysiology, 83

Prevention of CO poisoning 164-171
- at home, 164
- at work place, 165
- control of environmental pollution, 166
- in automobile and traffic, 164
- smoking cessation, 166-171

## S

Smoking
- and CO poisoning, 65
- cardiovascular effects, 71
- COHb levels, 66, 67
- passive, 68
- pregnancy, 73

Smoking cessation, 166-171
- clinical trials, 169-170
- general measures, 167-168
- transdermal nicotine, 168-170